STEM Education from Asia

Asia is the largest continent in the world. Five out of the top ten high performing economies in the Programme for International Student Assessment (PISA) 2018 are located in Asia. Why do Asian students perform so well in STEM-related subjects? This book answers this by examining the STEM education policies and initiatives in Asian economies, as well as the training programmes undertaken by STEM teachers in Asia.

The book is divided into four sections, each accompanied by a passage of commentary that summarizes the key takeaways of the chapters. Section one focuses on STEM policy environments and how various countries have developed policies that promote STEM as an integral part of national economic development. Section two focuses on STEM teacher education in the Philippines and Thailand, while section three focuses on STEM curriculum design, context, and challenges in four Asian economies. The fourth and final section focuses on presenting snapshots of STEM education research efforts in Malaysia, South Korea, and Singapore.

Written by Asian academics, this book will provide valuable insights to policy makers, educators, and researchers interested in the topic of STEM education, especially in the Asian context.

Tang Wee Teo is an Associate Professor at the National Institute of Education (NIE), Natural Sciences and Science Education (Academic Group) and Co-Head of the Multi-centric Education, Research and Industry STEM Centre at the NIE, Nanyang Technological University, Singapore. She has a particular interest in equity issues in STEM education. She is the Editor-in-Chief of the journal, Research in Integrated STEM Education.

Aik-Ling Tan is an Associate Professor at the National Institute of Education, Natural Sciences and Science Education (Academic Group), Nanyang Technological University, Singapore. Her research interest lies in students' learning in science, science teachers' professional development and curriculum design for integrated STEM education.

Paul Teng is the Dean and Managing Director at NIE International (NIEI), the education consulting company of Nanyang Technological University, Singapore. He is also an Adjunct Senior Fellow in the Natural Sciences and Science Education (Academic Group) at the National Institute of Education. His education research interests include pedagogies for adult learning in professional development.

Routledge Critical Studies in Asian Education
Series Editors: S. Gopinathan, Wing On Lee and
Jason Eng Thye Tan

For more information about this series, please visit: https://www.routledge.com/Routledge-Critical-Studies-in-Asian-Education/book-series/RCSAE

STEM Education from Asia
Trends and Perspectives

Edited by Tang Wee Teo, Aik-Ling Tan
and Paul Teng

Routledge
Taylor & Francis Group
LONDON AND NEW YORK

The
HEAD
Foundation

First published 2022
by Routledge
2 Park Square, Milton Park, Abingdon, Oxon OX14 4RN

and by Routledge
605 Third Avenue, New York, NY 10158

Routledge is an imprint of the Taylor & Francis Group, an informa business

British Library Cataloguing-in-Publication Data
A catalogue record for this book is available from the British Library

Library of Congress Cataloging-in-Publication Data
A catalog record has been requested for this book

ISBN: 978-0-367-56915-0 (hbk)
ISBN: 978-0-367-56916-7 (pbk)
ISBN: 978-1-003-09988-8 (ebk)

DOI: 10.4324/9781003099888

Typeset in Galliard
by KnowledgeWorks Global Ltd.

Contents

Figures

Tables

Introduction

Tang Wee Teo, Aik-Ling Tan, and Paul Teng

STEM is the acronym for science, technology, engineering, and mathematics. The genesis of the term embodies the political agenda of the United States to invest heavily in innovation through research and development in the STEM-related fields to improve the competitiveness of the nation. When the America Creating Opportunities to Meaningfully Promote Excellence in Technology, Education and Science (COMPETES) Act was signed into law on August 9, 2007, by former US President George W. Bush, it officiated the provision of large sums of US funding to promote STEM research, development, and education. This resulted in large numbers of STEM-specific programmes, committees, organisations, research centres/groups, universities, and laboratories being established. Since then, the global STEM wave has risen and Asia is not spared from the tsunamic effect of STEM.

Parallel to growing economic influence and affluence in many parts of Asia, there is increased investment in quality education to ensure that there is a sufficient pool of people skilled in STEM to sustain a nation's security and growth. Although STEM originated in the United States for a similar purpose, this chameleon has taken different shapes and forms in different parts of the world. Currently, the STEM education literature is dominated by works published mostly in the United States and Australia. Much less has been reported about how STEM education is understood, theorised, and studied in Asia.

Countries in Asia are undergoing dramatic structural transformations in their economies, especially away from their agrarian roots towards new phases that involve science, technology, and engineering, with increased interest and the need for knowledge in space science. The space race is made more acute with events such as the launch of the Chinese space shuttle (Long March 7) and commercial interest in space exploration. Unlike our Western counterparts, who are more cooperative in their space race, it appears that Asian nations are attempting to conquer space alone as it is linked to national prestige (Moltz, 2011). Consequently, this is likely to generate pressure on STEM education to provide the space industry elites. Furthermore, more Asian economies[1] are demonstrating outstanding performance in international benchmark tests such as PISA[2] and TIMSS[3], and policy makers, scholars, teachers, and other educators are interested to learn more about why students in these economies are learning better, what are some enabling factors to promote positive educational outcomes, and how other economies can emulate some best practices from these economies.

This is the first book with a focus on STEM education in Asia. It aims to draw the attention of individuals and groups interested to learn more about this topic from the voices in Asian economies. However, we are not claiming that the book covers *all* that is going on in Asia as it is the largest continent on Earth and the most socially, culturally, politically, and economically diverse.

As many parts of Asia begin to grow in affluence and more developing economies join in the economic race to the top, education has been recognised as one of the key forces to ensure that there is a critical pool of human resources to drive and sustain growth. This explains why many economies around the world are attracted to the rhetoric of harnessing STEM education to develop future generations of experts with integrated disciplinary knowledge and skills for "Mastering the Fourth Industrial Revolution" (theme of the 2015 World Economic Forum).

This book offers insights into STEM education policies, curriculum, teacher education, and education research in Asia for readers interested to learn more about how various Asian economies have internalised and integrated STEM in their unique contexts. One can find similarities in comparison to what has been reported in the United States, where the acronym "STEM" was coined, and other economies that have made a head start in STEM education. On the other hand, with a critical eye and situated experience, one will also notice the nuances in how STEM education is understood and played out within and beyond Asia. This book will contribute to a more inclusive discourse about STEM education worldwide but from a strong Asian perspective.

As STEM education continues to gain traction and morph into different forms reflective of the political, social, cultural, and educational valuing of the individual economies, readers should expect to read more about the fast-changing landscape of STEM education emerging from this part of the world. We strongly believe that this book shall be the first, but definitely not the last, of the voices emerging from STEM educators in Asia.

Notes

1. In the PISA 2018 tests, students in Asian economies ranked the top four places in Mathematics and top three places in Science (Source: https://www.oecd.org/pisa/PISA-results_ENGLISH.png).
2. PISA is the acronym for Programme for International Student Assessment, a worldwide study by the Organization for Economic Co-operation and Development (OECD).
3. TIMSS is the acronym for Trends in International Mathematics and Science Study is a large-scale international study conducted by the International Association for the Evaluation of Educational Achievement.

Reference

Moltz, J. C. (2011). Asia's space race. *Nature, 480,* 171–173.

Section 1

STEM policies

Rationale, imperatives, expected outcomes, and challenges

Paul Teng

This commentary examines the science, technology, engineering, and mathematics (STEM) policy environments in the context of regional and global trends, and then reviews the policies in Hong Kong, Malaysia, and Bhutan before making suggestions for the way forward. It is not the intention in this commentary to discuss what constitutes STEM education as this has been done elsewhere (Bybee, 2010) and in other parts of this book. Suffice to state that Wahono et al. (2020) have provided a broad interpretation that is used here – the teaching, learning, and integration of the four STEM disciplines (and skills) into STEM topics that emphasise real-world problems.

Introduction: Rationale and background behind STEM education policies

STEM policies around the world show almost single-mindedness about the purpose of STEM education, which is to prepare students for a variety of careers so that they can contribute to the economy, and to increase the overall scientific literacy of the population (Gough, 2014). This further appears to be a central preoccupation of policy makers, many of whom are likely influenced by the fear of losing ground in an increasingly technology-dominated world, as we can see from the papers on policy from Hong Kong (Chen Yu, Luo Tian, & So Wing Mui Winnie, Chapter 1), Malaysia (Lilia Halim, Lay Ah Nam, & Edy Hafizan Mohd Shahali, Chapter 2), and Bhutan (Kinley Kinley, Reeta Rai, & Sherab Chophel, Chapter 3).

Projections by the World Economic Forum (WEF) on the future of work and on the requirements for specific skills related to science and technology further spurred governments to examine their own policies towards STEM (World Economic Forum, 2020). Indeed, this latest report noted that the pace of technology adoption is expected to remain unabated and may accelerate in some areas. The adoption of cloud computing, big data, and e-commerce remain

DOI: 10.4324/9781003099888-1

high priorities for business leaders, following a trend established in previous years. However, there has also been a significant rise in interest for encryption, non-humanoid robots, and artificial intelligence. As the World's premier annual forum in which political and business leaders meet to discuss and debate on trends, the WEF has a strong influence on the human capital development policies of countries, and relatedly, on reforms in education policies that address the anticipated changes in the structure of economies and their requisite workforce.

In many respects, the concerns expressed by the WEF (2020) about the preparedness of students for the future workforce and future technology-strong work environment have led Hong Kong, Malaysia, and Bhutan to enact both legislation and policies to increase STEM education in their school systems, as elaborated in Chapters 1 to 3 of this book.

Policies are important catalysts and enablers for change in education systems worldwide. Having policies that promote STEM as an integral part of national economic development will therefore be a strong incentive to also strengthen STEM education. With STEM education, many Asian countries have enacted policies and supported them with implementation instruments, such as dedicated funding to accelerate the incorporation of STEM into existing school curriculum or even to develop new curricula (Wahono et al., 2020).

STEM policies generally (in contrast to STEM education policies) appear to have some common agenda items (Gough, 2014), such as:

- Supporting economic development through uplifting human capital in STEM qualifications so that graduates may be prepared for a wide range of occupations, professionally and vocationally (Panth, 2019);
- Enlarging the high-end STEM skilled workforce to engage in research and development leading to innovations;
- Improving the overall scientific literacy of the population; and
- Attracting more students to study STEM at secondary school and higher education levels.

A common dilemma in policy-making for STEM education is the lack of a common, agreed-upon understanding of what constitutes STEM and how schools should implement a STEM curriculum (Quraan & Forawi, 2019). This is reflected in curriculum design and the relative emphasis put on different subject matters within the scope of STEM, as shown in our comparison of the three geographies in Chapters 1, 2, and 3.

Policies on STEM education are not new to Asia. As noted by Chen et al. (Chapter 1), the Hong Kong government has invested significant resources between 2015 and 2020 to support its policy imperative of STEM as the basis for maintaining innovation. In Malaysia, the precursor to STEM is science education (Chapter 2). The 60:40 Science/Technical: Arts (60:40) Policy in education, established in 1967 by the Malaysia Higher Education Planning Committee, was enforced in 1970 and targets 60% among all upper high school students concentrating on science while 40% participating in arts. Even in

Bhutan, science education started from the adoption of Western education in 1961 (Chapter 3), and the government's 10th Five Year Plan (2008–2012) prioritised the reform of the science curriculum and conducted a needs assessment study. A STEM curriculum was subsequently put in place, to adopt an interdisciplinary and applied approach to STEM, which today is identified as one of the essential learning areas in the Bhutanese curriculum to provide innovative and creative skills under the science and technology education required for the 21st century.

In retrospect, the issues surrounding STEM education are not that different from the issues that have plagued science education policy makers, in particular, the issue as to what are the educational purposes that science and technology education can best provide for students as they move through the stages of schooling (Fensham, 2008). This seems to be the case with Bhutan, a country that is seeking to balance its traditional strengths with the need to modernise.

Policy imperatives and instruments

Although many countries have developed policies to support STEM education (Gough, 2014), such as the over-arching policies in Hong Kong, Malaysia, and Bhutan, far fewer have specific policy instruments that support implementation. Experience has shown that generally, policies on their own are just articulations of intent by governments and are destined to be ineffective unless backed by instruments such as guidelines, statutes, measures, or directives that are funded for implementation (Panth, 2019). Indeed, adequate funding and coordination have been identified elsewhere as important imperatives to accompany espoused STEM policies (Education Commission of the States, 2016).

Supporting measures to reinforce policies

How has policy been supported by dedicated resourcing in Hong Kong, Malaysia, and Bhutan?

Hong Kong has shown multiple measures to boost STEM in the context of innovation and technology. As discussed in Chapter 1, an Innovation and Technology Bureau was set up in 2015, and the Government's Innovation and Technology Fund (ITF) had provided approximately HKD 8.9 billion for more than 4,200 projects by that year. Multiple funds (exemplified by the Innovation and Technology Venture Fund worth HKD 2 billion) provided the impetus to make STEM visible in the minds of students and the populace. The Hong Kong government further spurred the development of innovation and technology start-ups by providing incentives so that a private sector requiring STEM expertise would feedback to STEM education.

Malaysia actioned its STEM policies through the Malaysia Education Blueprint 2013–2025 (MEB 2013–2025) which is the National Education Blueprint that was aimed at supporting Malaysia's vision of a first-class education within the Malaysian socio-culture context. One of the strategic thrusts in the MEB is to increase the quality of STEM education. This led to the STEM education initiative with three core strategies – (1) Raising students' interest in STEM education, (2) Upskilling teachers' competencies, and (3) Enhancing parents' and students' awareness toward STEM. Through this initiative, resources were allocated by the government through the Ministry of Education to develop instructional resources for STEM education, as well as address teacher professional development. Towards this end, a National STEM Centre (Pusat STEM Negara) was established in 2018. The initiative has also seen many STEM outreach programmes to promote students' and public awareness of STEM and STEM education.

Bhutan is a relative newcomer with respect to introducing measures to intensify STEM education in its K-12 classrooms. It has chosen to recognise Information Communication Technology (ICT) as an important tool to achieve its development objectives, and additionally to improve its e-connectivity for education (Chapter 3). Its policy imperative is strongly based on using ICT as an enabler and a goal for STEM education and has invested to do so. A justification which is interesting is that its dispersed geography with contrasting contours (mountains, valleys, and plains) argues strongly for e-connectivity in education.

Role of teachers

Many studies have shown the importance of teachers who are capable of implementing a STEM curriculum. Teacher professional development was identified by Quraan and Forwai (2019) as a recurring theme in the studies they reviewed, especially in areas beyond just the content associated with STEM such as collaboration, authentic learning, integration of real-world experiences into STEM classroom practices.

In a comprehensive study, Fensham (2008) noted the importance of developing a STEM-capable teaching force and recommended that (1) Policy makers should consider the policy implications (financially and structurally) and the benefits in establishing the provision of ongoing, focused professional development in science and technology and their teaching, as an essential aspect in the careers of all science teachers, and (2) Policy makers should consider, within whatever funding is available, how to maximise the number of students whose science and technology education is in the hands of able science teachers. In all three geographies (Hong Kong, Malaysia, and Bhutan) reviewed in Chapters 1 to 3 in this book, the role and associated capability of teachers have been noted as important for the success of STEM education policies and their mandated curricula, and rightly so.

Attracting students into STEM

There also appears to be tension between government policies advocating for more STEM education and students' declining interest in studying STEM subjects (Gough, 2014). However, this does not seem to be the case in Hong Kong, as the plethora of policies and efforts are, to a large extent, helping to create a supportive policy context for STEM education in Hong Kong, contributing to the flourishing of STEM learning activities both in and outside of schools, and augmented by an economy that is built on innovation and technology (Chapter 1).

In Malaysia, amidst rapid growth in STEM-related industries (Chapter 2), the number of students taking STEM-related subjects had decreased yearly and was occurring at all levels of education in Malaysia. Despite the demand for STEM-related occupations, which were the top emerging jobs, the STEM skilled workforce was on the decline. The Malaysian Ministry of Education (MOE) reported that in 2018, in upper secondary, only 83,608 (22%) of Malaysian students were in the STEM stream as compared to 92,956 (24%) in 2017, with a reduction of 9,348 students (MOE, 2018). In their paper (Chapter 2), factors contributing to this decline were suggested to be (1) Limited awareness about the value of STEM learning and careers among students and parents, (2) Perceived difficulty of STEM subjects, (3) Content-loaded curriculum with less emphasis on relevance to everyday life and opportunities to be creative and innovative, (4) Teaching and learning approaches are teacher-centred, and (5) Limited and outdated infrastructure (Chapter 2).

Bhutan does not currently have a noteworthy technology sector in its economy and there does not appear to be any evidence to show any decline in interest to pursue STEM subjects or STEM careers (Chapter 3).

It is of interest to note that the development of STEM education in the Western hemisphere was motivated by the low interest of the younger generation in work related to the STEM disciplines (Chesky & Wolfmeyer, 2015). Apart from the three geographies compared in this book, this situation has also been observed in other Asian countries even as economies in the region have increasingly moved towards science and technology-enabled activities (Wahono et al., 2020).

Instructional policies

In their extensive review, Quraan and Forawi (2019) identified themes that were of concern to teachers involved in STEM education – subject integration, STEM content, professional development, time, assessment, outside support, collaboration, willingness, authentic experiences, leadership, and dissent – which they suggested needed to be addressed through policies directed at them. The implication is that unless there are purposive policies, there is little incentive for teachers to recognise them.

In his study, Fensham (2008) also recommended that policy makers should consider a quite different curriculum for science and technology in the primary years, that engages considerable pedagogical skills of the teachers, provides the young learners with a series of positive and creative encounters with natural and human-made phenomena and builds their interest in these two areas of learning (Fensham, 2008).

Both Hong Kong and Malaysia appear to have purposeful policies aimed at instructing STEM, as discussed in Chapters 1 and 3. In Bhutan, the current Bhutanese education system provides segregated domain-specific STEM education. From the acronym STEM, only science (S) and mathematics (M) and Information and Communications Technology (T) are mandatory core curriculum of the Bhutanese Education system (BES) from primary to middle secondary level (Grade X) (Chapter 3).

STEM policies: Matching expected and real outcomes

In the introduction section of this commentary, I noted that the WEF has a strong influence on the human capital development policies of countries, and relatedly, on reforms in education policies that address the anticipated changes in the structure of economies and their requisite workforce. The common STEM policy outcome is generally aimed at providing a "future-ready" workforce for economies at different stages of development, for example, to enable continued economic competitiveness (Hong Kong), strengthen the human capital base for economic growth (Malaysia), and develop a base for modernising the economy (Bhutan), respectively covered in Chapters 1 to 3.

A pointed question that has to be asked is whether the learning outcomes from STEM curricula do indeed meet the anticipated demands of the future workforce and economy. As noted by Wahono et al. (2020), the anticipated learning outcomes from enacting STEM policies and practices focus on academic learning achievements, higher-order thinking skills, and motivation toward one or more of the STEM disciplines, and eventually a career in a STEM-based vocation.

There has also been an expectation that STEM enactment will lead to the inculcation of 21st-century competencies (the so-called "21CC") in participating students and empower them with soft skills in future new economies (Teng, 2019; WEF, 2020). These represent the higher-order thinking skills such as critical thinking, creativity, collaboration, etc. As yet, there does not seem to have been any comprehensive assessment as to whether this has been achieved in Hong Kong, Malaysia, or Bhutan, although in the latter it may be early days yet for policy imperatives to have exerted any impact.

As noted earlier, one of the two goals of STEM education is to support economic development (Gough, 2014). Hong Kong seems to have been successful in navigating the space between STEM education and the need for an industrial STEM workforce, as noted in Chapter 1. In Chapter 2 on Malaysia, uncertainty is expressed as it appears that there is a lack of quantitative data to show if government policies and their implementation have met the needs of an evolving economy. Likewise in Bhutan, outcomes from curriculum renewals are yet to show their full potential in supporting a modern economy in which technology is important (Chapter 3).

The match between STEM student outcomes and the needs of the economy have been warned by Metcalf (2010) to be strongly influenced by the "linear

STEM pipeline model", in which supply-side projections of manpower numbers over-influence policy decisions and under-consider aspects of culture and gender. For the Asian region, countries need to be continually sensitive to a possible mismatch between how many STEM graduates are produced by the education system versus the needs of evolving economies (Panth, 2019).

With respect to the other goal of STEM education, to produce a science and technology literate population as an outcome, it is not possible to assess if this has been achieved in Hong Kong, Malaysia, and Bhutan, the three geographies presented in this book. Hong Kong is likely the closest as the populace, public, and private sectors seem to collectively show strong support for a strong STEM base to generate an economy with strong capabilities in innovating technology.

Challenges and way forward

Efforts at promoting STEM education to students and attracting more interest in STEM careers may be stymied if socio-cultural attributes of the populace are not taken into consideration, including gender issues. In his report, Fensham (2008) recommended that policy makers should consider means of overcoming cultural disadvantages that some groups of students experience specifically in science and technology education.

Across the world, STEM policies appear to have some common agenda items (Gough, 2014), and Asia is no different:

- Supporting economic development through uplifting human capital in STEM qualifications so that graduates may be prepared for a wide range of occupations, professionally and vocationally;
- Enlarging the high-end STEM skilled workforce to engage in research and development leading to innovations;
- Improving the overall scientific literacy of the population; and
- Attracting more students to study STEM at secondary school and higher education levels.

Each of these agenda items demands a different response depending on the geography concerned. Although many countries have made STEM education a national priority (e.g., the United States of America (USA)), much remains to be done to encourage students to go into STEM and subsequently for society and the economy to reward those embarking on STEM careers (STEM Education Coalition, 2019).

In the United States of America, states like Utah have enacted statutes to foster coordination of STEM education at different levels from K-12 to higher education, and have done so by establishing STEM Coordination Centres (Education Commission of the States, 2016). At a national level, countries like the United States of America have recognised the importance of linking K-12 STEM education with STEM-related programmes in institutes of higher learning, so that there are observable academic benefits from taking STEM subjects in school when applying for higher education. In looking forward, can Asia learn from the experiences of Western countries which have had a long history of technology-enabled economic growth?

Ensuring the relevance of STEM and the "suitability to purpose" for employment is an issue that should continue to be at the forefront of policy makers. As Gough (2014) rightly noted, there is a "muddy" relationship between STEM qualifications, industry demand, and employment or employability. This becomes even more disconnected if the different government agencies responsible for setting science education policy, projecting manpower needs, liaising with the private sector (industry), and conducting strategic planning for the economy, are disconnected and not aligned with each other.

Even when government efforts are successful in aligning the economy's manpower needs with the products of its education systems, a remaining question is that of raising scientific literacy in the general population (Gough, 2014). At a time in human history when many of the pressing issues of the day such as climate change, declining natural resources, food security, and generally adverse anthropogenic effects on the natural world require some modicum of scientific knowledge to make informed decisions on participation, we are also seeing a small but anti-science movement in some countries which may derail the best intentions of STEM policies. Policy makers should consider replacing the generic use of "scientific literacy", as a goal of school science education, with more precisely defined scientific knowledge and scientific abilities, that have meaning beyond school for the students at each of the stages of schooling, for example, lower primary, upper primary, lower secondary, the last years of compulsory schooling, and the final secondary years (Fensham, 2008).

Moving forward, there is general agreement that STEM education and the role of STEM in economic development needs to be supported. This is evidenced by the developments in Hong Kong, Malaysia, and Bhutan reviewed in this section of the book. As noted earlier, STEM education has been integrated into policies in Hong Kong from 2015 to 2020 (Chapter 1). Today, innovation and technology and STEM education are advocated as policy-driven initiatives in Hong Kong. The Hong Kong Education Bureau has been stepping up efforts to promote STEM education, with ongoing renewal of the school curricula and professional training for STEM teachers. In Malaysia, to strengthen the country's STEM ecosystem, the National STEM Strategic Action Plan 2018–2025 has been developed. In Bhutan, STEM education was bolstered by the 2020 *kasho* (royal edict) for education reform in which the King of Bhutan encouraged new and revitalised educational practices to prepare students for the modern era; STEM education has become a key priority for the education system in Bhutan (Chapter 3).

It would, therefore, appear that the policy commitments are strong in Hong Kong, Malaysia, and Bhutan to make STEM an integral part of national development in the near future.

References

Bybee, R.W. (2010). What is STEM Education? *Science, 329*(5995), 996.

Chesky, N.Z. & Wolfmeyer, M.R. (2015). *Philosophy of STEM education: A critical investigation*. New York: Palgrave Macmillan.

Education Commission of the States (2016). *A State Policymaker's STEM Playbook*. Accessed 24 May 2021. https://www.ecs.org/a-state-policymakers-stem-playbook/

Fensham, P.J. (2008). *Science education policy-making: Eleven emerging issues*. Paris: UNESCO. Accessed 25 May 2021. http://efepereth.wdfiles.com/local–files/science-education/Science_Education_Policy-making.pdf

Gough, A. (2014). STEM policy and science education: Scientistic curriculum and sociopolitical silences. *Cultural Studies of Science Education, 10*(2). doi: 10.1007/s11422-014-9590-3.

Metcalf, H. (2010). Stuck in the pipeline: A critical review of STEM workforce literature. *UCLA Journal of Education and Information Studies, 6*(2). Accessed 25 May 2021. https://escholarship.org/uc/item/6zf09176

Ministry of Education (MOE). (2018). *Quick fact 2018: Malaysia educational statistics*. Putrajaya: Educational Planning and Research Division, Ministry of Education Malaysia. Accessed on 20 May 2021. https://www.moe.gov.my/penerbitan/1587-quick-facts-2018-malaysia-educational-statistics-1/file

Panth, B. (2019). STEM education to drive learning and innovation. In *STEM Education: An Overview. THF Workshop Reports No. 7* (pp. 43–52). Singapore: The Head Foundation. https://headfoundation.org/2019/11/17/workshop-report-no-7/

Quraan, E.J.A. & Forawi, S.A. (2019). Critical analysis of international STEM education policy themes. *Journal of Education and Human Development, 8*(2), 82–98.

STEM Education Coalition. (2019). *The case for STEM Education as a National Priority: Good Jobs and American Competitiveness*. Accessed 25 May 2021. http://www.stemedcoalition.org/wp-content/uploads/2019/10/Sept-2019-Fact-Sheet-PDF-STEM-Education-Good-Jobs-and-American-Competitiveness.pdf

Teng, P. (2019). What is STEM education? In *STEM education: An overview. THF Workshop Reports No. 7* (pp. 9–14). Singapore: The Head Foundation. https://headfoundation.org/2019/11/17/workshop-report-no-7/

Wahono, B., Lin, P.L. & Chang, C.Y. (2020). Evidence of STEM enactment effectiveness in Asian student learning outcomes. *International Journal of STEM Education, 7*. https://doi.org/10.1186/s40594-020-00236-1/

World Economic Forum. (2020). *The Future of Jobs Report 2020*. Accessed 25 May 2021. http://www3.weforum.org/docs/WEF_Future_of_Jobs_2020.pdf

1 STEM education policies in Hong Kong

Yu Chen, Tian Luo, and Winnie Wing Mui So

Introduction

Hong Kong is fast becoming known as a global innovation and technology hub. In the 2020 edition of the Global Innovation Index, Hong Kong's innovation and technology sector ranked second in the world (Dutta et al., 2020). As a means of making people innovative and productive in all of their endeavours, Science, Technology, Engineering, and Mathematics (STEM) education implemented in schools, different organisations, tertiary institutions, and private companies is arguably important for innovation and technology. This chapter aims to provide a comprehensive review of the STEM policies in Hong Kong and their implementation by key stakeholder groups (i.e., schools, universities, and other organisations). To be specific (Figure 1.1), the first section discusses the government's policies for innovation & technology and STEM education, followed by the STEM initiatives of the Education Bureau (EDB), which are illustrated in the second section. The third section describes the implementation of STEM education policies by key stakeholder groups.

Government STEM policies

This section presents the policy emphasis on promoting innovation and technology and STEM education in Hong Kong in accordance with the government's Policy Addresses in the past few years from 2015 to 2020. It could be observed that the government provided substantial financial, technical, and human resources in an effort to vigorously promote innovation and technology in Hong Kong. The government has also paid great attention to enhancing the cooperation between the Mainland and Hong Kong to further benefit the development of local innovation and technology. Similarly, the government has invested significant resources in promoting STEM education in primary and secondary schools. More details about STEM education policies in terms of boosting innovation & technology and strengthening STEM education are provided as follows.

DOI: 10.4324/9781003099888-1

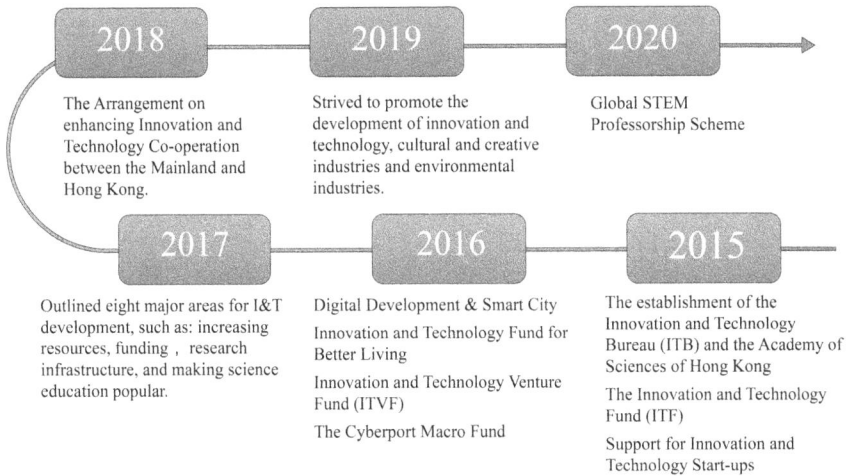

2018	2019	2020
The Arrangement on enhancing Innovation and Technology Co-operation between the Mainland and Hong Kong.	Strived to promote the development of innovation and technology, cultural and creative industries and environmental industries.	Global STEM Professorship Scheme

2017	2016	2015
Outlined eight major areas for I&T development, such as: increasing resources, funding , research infrastructure, and making science education popular.	Digital Development & Smart City Innovation and Technology Fund for Better Living Innovation and Technology Venture Fund (ITVF) The Cyberport Macro Fund	The establishment of the Innovation and Technology Bureau (ITB) and the Academy of Sciences of Hong Kong The Innovation and Technology Fund (ITF) Support for Innovation and Technology Start-ups

Figure 1.1 Policies on innovation & technology (cited from Policy Address 2015–2020).

Boosting innovation & technology

The Hong Kong government's efforts to boost and strengthen the innovation and technology sector can be traced back to the 1990s. In the 1997 Policy Address, adapting to new technologies and developing new industries were emphasised as always being important for Hong Kong. The importance of innovation and technology has been more recognised in the Chief Executives' Policy Addresses since 2015. Figure 1.1 summarises the important measures for innovation and technology from 2015 to 2020.

The government, in its 2015 Policy Address, explained the important role of innovation and technology in economic development: "Innovation and Technology can diversify the economy, provide wider employment opportunities in research and development (R&D), and enhance the competitiveness and growth of related industries" (The Government of the Hong Kong Special Administrative Region, 2015, p. 15). The next year's Policy Address (The Government of the Hong Kong Special Administrative Region, 2016) further elaborated that innovation and technology are important factors affecting both economic growth and people's daily life: "Innovation and technological capabilities are key indicators of the level of social and economic development. Innovation and technology not only create new momentum for economic growth, they also provide us with a more convenient, comfortable and secure way of living" (p. 20). The government again pledged to promote innovation and technology to "foster economic development and improve people's daily lives" (The Government of the Hong Kong Special Administrative Region, 2017, p. 23).

Multiple measures have been adopted to boost innovation and technology in Hong Kong. In 2015, an Innovation and Technology Bureau was set up, and the Government's Innovation and Technology Fund (ITF) had provided approximately HKD 8.9 billion for more than 4,200 projects by that year (The Government of the Hong Kong Special Administrative Region, 2015). After that, the 2016 Policy Address pointed out more measures for Hong Kong's innovation and technology. For instance, multiple funds such as the Cyberport Macro Fund (HKD 200 million), the Innovation and Technology Venture Fund (HKD 2 billion), as well as the Innovation and Technology Fund for Better Living (HKD 500 million) were launched (The Government of the Hong Kong Special Administrative Region, 2016). Moreover, continued full support was provided to innovation and technology start-ups, such as increasing the incubation scheme quotas or the provision of facilities and spaces to cater to the needs of the development of industry. To build Hong Kong into a world-class smart city, the government proposed expanding the coverage of free Wi-Fi services, developing more user-friendly mobile applications (apps) for the public, and formulating policies based on big data analysis (The Government of the Hong Kong Special Administrative Region, 2016).

The 2017 Policy Address also included a significant portion of innovation and technology. In addition to doubling the Gross Domestic Expenditure on Research & Development to approximately HKD 45 billion a year, the government outlined seven other major areas for innovation and technology (I&T) development, including the cultivation of technology talents, collaboration with overseas scientific research institutions, the provision of investment funding, the provision of research infrastructure, the review of existing legislation and regulations, the opening up of government data, as well as popular science education (The Government of the Hong Kong Special Administrative Region, 2017). In the 2018 Policy Address, the Arrangement on Enhancing Innovation and Technology Co-operation between the Mainland and Hong Kong which provides an overarching framework for mutual I&T collaboration was signed. An HKD 1 billion Construction Innovation and Technology Fund was established, and a new Research Matching Grant Scheme with a total commitment of HKD 3 billion to boost the research and development work of universities was launched, together with HKD 500 million allocated for supporting the further development of the smart city (The Government of the Hong Kong Special Administrative Region, 2018).

In the 2019 Policy Address, the stated aim was to "strive to promote the development of innovation and technology, cultural and creative industries and environmental industries" (The Government of the Hong Kong Special Administrative Region, 2019a, p. 25), suggesting the government's continued determination to promote innovation and technology. The more recent 2020 Policy Address (The Government of the Hong Kong Special Administrative Region, 2020a) proposed launching a Global STEM Professorship Scheme at an estimated cost of HKD 2 billion to attract more talents overseas to engage in I&T related teaching and research activities in Hong Kong. The objective is to build up the capacities of universities in commercialising and applying their research discoveries and encourage technology and knowledge transfer, and eventually spearhead I&T development in Hong Kong in the long run.

Strengthening STEM education

STEM education has become a highlight of the Chief Executives' Policy Addresses to achieve innovation and technology development. Key highlights related to STEM education in the Policy Addresses during 2015 to 2020 are listed in Figure 1.2.

In the 2015 Policy Address, the government asked the EDB to improve the curricula and learning activities of the STEM education Key Learning Areas (KLAs) and to enhance the training of teachers to fully unlock the potential for innovation among primary and secondary school students (The Government of the Hong Kong Special Administrative Region, 2015). The 2016 Policy Address clearly stated that the Hong Kong government will make efforts to promote STEM and students' pursuit of studying these subjects (The Government of the Hong Kong Special Administrative Region, 2016).

A great deal of effort and progress was made in 2017. More detailed measures were proposed, such as expanding the Internship Programme to STEM graduates, enhancing the professional development of STEM teachers, and setting up a STEM Education Centre (The Government of the Hong Kong Special Administrative Region, 2017). The STEM Education Centre supported by the EDB provides primary and secondary school teachers with teacher professional programmes as well as relevant support for learning and teaching. Moreover, schools can apply for the funding of the Quality Education Fund for schools to enhance or diversify their school-based STEM curriculum. The 2017 Policy Address suggested that the EDB would strive to promote STEM education. Therefore, apart from above support, the EDB had released the One-off STEM Grant to support STEM education in primary and secondary schools during the 2015/2016 to 2018/2019 school years.

For the 2019/2020 school year, the 2018 Policy Address launched a new recurrent Life-wide Learning Grant with an annual provision of HKD 900 million to benefit students' out-of-classroom experiential learning activities, including those related to STEM education (The Government of the Hong Kong Special Administrative Region, 2018). The aim was to "continue to enhance

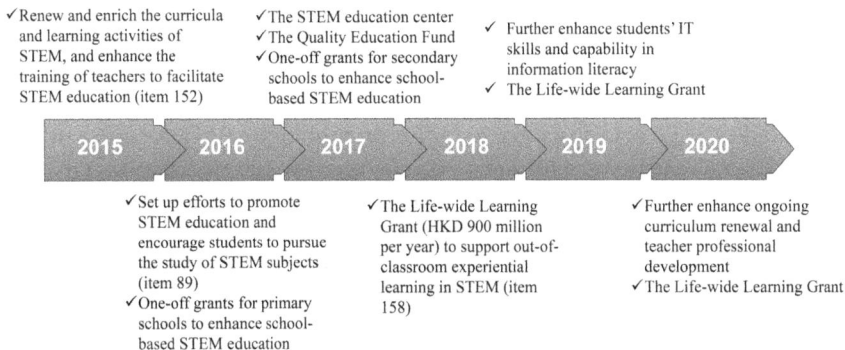

Figure 1.2 Policies on STEM education (cited from Policy Address 2015–2020).

the support for schools in providing students with more learning, exchange and competition opportunities, with a view to unleashing their potentials in science and technology" (p. 35).

The 2019–2020 Policy Address further stressed the sustained efforts needed to be made by the EDB in advancing STEM education. The Legislative Council Panel on Education 2021 Policy Address (The Government of the Hong Kong Special Administrative Region, 2019b) highlighted that building on the existing foundation, the EDB would aim to further enhance students' IT skills and their capability in information literacy, through providing a series of teacher professional development programmes and a variety of learning and teaching resources for teachers, as well as offering the Life-wide Learning Grant and the funding of the Quality Education Fund for schools. Similarly, in the Legislative Council Panel on Education 2020 Policy Address (The Government of the Hong Kong Special Administrative Region, 2020b), it was set out that sustained efforts by the EDB would be made to further enhance ongoing curriculum renewal, or launch more teacher professional development programmes, in order to further promote the progressive implementation of STEM education in schools. Schools can continue to use the Life-wide Learning Grant and the Quality Education Fund to enhance school-based STEM education.

STEM initiatives by the Education Bureau

To prepare students well for the local, social, and global changes in a knowledge-based, technologically advanced, and increasingly globalised world, a series of initiatives for STEM education have been set up by the EDB, including the release of the report, "Promotion of STEM Education – Unleashing Potential in Innovation" in 2016; updates to the Curriculum Guides of Science, Technology, and Mathematics Education in KLAs and the primary General Studies curriculum in 2017 and beyond; the release of "Computational Thinking – Coding Education: Supplement to the Primary Curriculum" in 2017; the Task Force on the Review of the School Curriculum from 2017 to 2020, as well as the Initial Training Programme and resources for STEM learning as support measures for teachers since 2017.

Releasing the report of promotion of STEM education

In 2015, the EDB carried out a series of consultation sessions with relevant stakeholders (e.g., principals, teachers) for promoting STEM education in different KLAs. Based on the information collected through the consultation, recommendations that pave the way for the promotion of STEM education were highlighted in the report of "Promotion of STEM Education – Unleashing Potential in Innovation" which was issued in 2016 in order to nurture "versatile talents with different sets and levels of knowledge and skills so as to enhance the international competitiveness of Hong Kong, and in turn contribute to national developments" (Education Bureau, 2016, p. 1). Six strategies for enhancing

STEM education were proposed in this report: (1) To renew the curricula of science, technology, and mathematics education KLAs; (2) To enrich learning activities for students; (3) To provide learning and teaching resources; (4) To enhance professional development of schools and teachers; (5) To strengthen partnerships with community key players; as well as (6) To conduct reviews and disseminate good practices (Education Bureau, 2016).

Integrating STEM in renewed curriculum guides

After the Learning to Learn Curriculum Reform in 2001, the EDB moved to a new phase of ongoing curriculum renewal – focusing, deepening, and sustaining given the need to maintain the next generations' competitiveness in the face of the massive changes globally in economy, science, technology, and society. In this regard, curriculum guides and their supplements of various KLAs, covering STEM-related subjects such as mathematics education, science education, and technology education were updated in 2017 and beyond, with the aim of nurturing students' creativity, collaboration, problem-solving skills, and innovativeness.

During the ongoing curriculum renewal, a great deal of attention has been paid to the integration of STEM education, with emphases on inquiry-based learning, integration and the application of knowledge and skills to solve authentic problems, and hands-on and minds-on activities. For instance, it is advised in the curriculum guide for science education (Curriculum Development Council [CDC], 2017a) that science education should provide learning experiences for students to comprehend the interdisciplinary nature of STEM, as well as to develop STEM skills and attitudes. According to the curriculum guide of mathematics education, two approaches to strengthening STEM education were recommended, of which one is to organise learning activities based on a topic in mathematics, and the second is to carry out project-based learning (CDC, 2017c). Moreover, technology education is suggested to partly aim at promoting the development of knowledge and skills among students to meet the challenges of the rapidly developing technologies in society. A brief summary of the guidelines for STEM education in the curriculum guides is presented in Table 1.1.

Promoting coding education in primary schools

Coding education has been promoted by the EDB as optional teaching content or extra-curricular activities in local junior and senior secondary schools since 2014 (Education Bureau, 2014). It aims at enabling students to "master the coding skills and apply the skills to different contexts to complete assigned tasks/jobs" (CDC, 2020, p. 3). Hong Kong primary schools are therefore recommended to cover the learning content (e.g., Automation, or Data Processing, Coding Concept and Practices) listed in "Computational Thinking – Coding Education" to cultivate students' computational thinking through implementing school-based programmes or theme-based teaching. To date, several local

Table 1.1 Elements of STEM education in the renewed curriculum guides in 2017

Curriculum guides	Emphasis on STEM education
General Education (Primary schools)	"The planning of STEM activities should: enliven students' interest in and curiosity about science and technology; strengthen students' understanding of everyday life matters and develop their abilities to integrate and apply knowledge and skills; enhance students' perseverance and abilities to make decisions, and to face challenges and solve problems with integrity; and to cultivate students' creativity and innovative spirit by reinforcing the design cycle in inquiry learning" (CDC, 2017d, pp. 79–80).
Science Education (Secondary schools)	"Science education provides learning experiences for students to develop scientific literacy with a firm foundation in science, realise the relationship between science, technology, engineering and mathematics, master the integration and application of knowledge and skills within and across KLAs, and develop positive values and attitudes for personal development and for contributing to a scientific and technological world" (CDC, 2017a, p. 19).
Mathematics Education (Secondary schools)	"Learning activities based on a topic in Mathematics for students to integrate relevant learning elements from other key learning areas (such as representing and comparing sizes of cells using scientific notation)" (CDC, 2017c, p. 43) "Projects for students to integrate relevant learning elements from different key learning areas" (CDC, 2017c, p. 44).
Technology Education (Secondary schools)	"[D]eveloping among students a solid knowledge base and enhancing their interest in technology for future specialisation studies and careers; strengthening students' ability to integrate and apply knowledge and skills (including skills related to hands-on experiences) within and across the KLAs of Science, Technology and Mathematics Education; fostering innovation in meeting the challenges of economic and technological development; strengthening the collaboration among teachers in schools and the partnerships with community stakeholders" (CDC, 2017b, p. 49).

schools have shown their interest in promoting coding education as a part of their regular school curriculum (Wong et al., 2015).

So, Jong, et al. (2020) proposed that coding education is considered to play an important role in nurturing students with adequate computational thinking skills. Computational thinking in practice refers to a problem-solving process including abstraction, decomposition, algorithmic design, evaluation, and making generalisations (Selby & Woolard, 2013), and is considered as essential in supporting students' development of skills related to problem solving, abstract thinking, and logical reasoning (Lye & Koh, 2014). Teaching coding to young students shows benefits, including enabling them to develop skills or attitudes to programming at the age when they can learn to code quickly (Duncan et al., 2014).

In order to promote coding education in primary school, a draft document called "Computational Thinking – Coding Education: Supplement to the Primary Curriculum" was published by the EDB in 2017. The formal document, "Computational Thinking – Coding Education: Supplement to the Primary Curriculum" was released recently. According to the EDB Circular Memorandum No. 108/2020[1], the formal document further clarifies the learning elements of Coding Education at upper primary level, mainly through enriching the learning content of "Connecting Computing", "Applications and Impacts", and "Computational Thinking Practices". To better connect the implementation of STEM education, the learning content of "Forming a System Connected with Physical Objects" has been refined. Through Coding Education, students are expected to be able to: "understand the basic computational thinking concepts and practices", "possess the ability to develop programs and process data to solve problems", "understand the process of problem solving and limitations of coding", "connect coding with real-life problems and other subjects", and "solve problems through communication and collaboration" (CDC, 2020, p. 5).

Yet, the connection of coding education to STEM education is not clearly articulated in all these government documents. Moreover, there have been studies reporting the disadvantages of teaching coding to primary school students. For instance, students may show negative attitudes toward coding if they are taught by teachers who lack confidence or skills in coding (Duncan et al., 2014). The successful inclusion of computational thinking in classrooms also depends on a diverse set of factors other than just policy priority. These factors include teacher training and confidence, learning tools and resources, how much students develop concepts, and how students are supported to continue coding study in their future (Duncan et al., 2014; Weintrop et al., 2016). Research is needed to provide evidence to support the effectiveness of coding education in students' STEM learning.

The task force on the review of the school curriculum

In 2017, the EDB organised the Task Force on the Review of the School Curriculum to holistically review the school curricula in primary and secondary schools. A consultation document of the Task Force released in 2019 formulated six directions of initial recommendations, including whole-person development, values education, creating space and catering for learner diversity, applied learning, university admissions, and STEM education. In particular, the consultation document recommended that the EDB give clearer definitions of STEM education to address the different needs of schools by stepping up territory-wide support for STEM education and to establish a designated committee to oversee the long-term development of STEM education in order to strengthen STEM education in primary and secondary schools.

Guided by the consultation document, the Task Force then carried out three months of consultation work with key stakeholder groups and veterans in the education field from June to October in 2019. Based on the views and comments collected, a final report of the Task Force, namely "Optimise the curriculum for the future: foster whole-people development and diverse talents", was submitted to the Government in 2020, which consolidated the final directional recommendations. For strengthening STEM education, in particular, continued efforts should be made on clearly defining STEM education and setting up the designated committee for the long-term benefits of STEM education. In addition to this, efforts should be put into the provision of school-based examples of STEM education, the STEM-related professional development of school teachers, the exploration of schools as regional STEM resources centres, as well as STEM education for gifted students (EDB, 2020).

Intensive Training Programme for STEM teachers

In parallel with the school curriculum renewal, the EDB has also been progressively organising symposiums and professional development for teachers involved in STEM education. The EDB started an Intensive Training Programme (ITP)[2] on STEM education from the 2017/2018 school year. To further enhance the professional development for STEM coordinators and frontline teachers, the EDB plans to conduct a new round of ITP from the 2020/2021 to the 2022/2023 school years. The aim of the ITP is to enhance the capacities of curriculum leaders and middle managers of all primary and secondary schools in planning and implementing school-based STEM-related activities and in holistic planning and cross-disciplinary collaboration.

Generally, four teachers from each school are invited to jointly take part in the ITP's seminars, workshops, and experience sharing sessions. Teachers will be engaged in curriculum planning, activity design, and the use of relevant learning, teaching, and assessment strategies relating to STEM education. Specifically, teachers will be provided with opportunities to develop or modify their schools' STEM curriculum, to enhance strategies of learning, teaching, and assessment, to update knowledge about the relevant STEM advancement, as well as to share experiences of STEM education among schools.[3]

Moreover, the EDB is continuing to develop learning and teaching resources to assist teachers' implementation of STEM education. For instance, teachers in primary and secondary schools are able to freely get STEM learning and teaching resources at the website of STEM education[4] or the STEM Education Centre[5]. To more effectively facilitate primary school students' learning in coding education, teachers can make reference to the eight learning modules suggested in the Modular Computer Awareness Programme[6], including: (1) Joy to the Computer World, (2) Drawing with a Computer, (3) Writing with a Computer and Word Processing, (4) Using the Internet, (5) IT Applications and Implications, (6) Calculation and Charting with Spreadsheets, (7) Using E-mail, and (8) Coding Education. However, none of the above materials match well with the integrated nature of STEM education.

Implementation of STEM education policies by key stakeholder groups

Given the STEM policies and initiatives, a majority of local schools have implemented STEM education so far. There are also many associations collaboratively promoting STEM education. With their joint efforts, a series of activities supporting STEM education have been successfully held.

Universities always play an important role in supporting teacher professional development and STEM education in schools. This section describes in detail how those key stakeholder groups, including teachers, schools, universities, and other associations, make efforts to enhance STEM education and to achieve the government's STEM policies.

STEM opportunities in primary and secondary schools

STEM education is mainly promoted in science, technology, and mathematics education (EDB, 2016). Despite the fact that the Hong Kong government has invested a large number of resources, the implementation of STEM education tends to be "school-oriented". As suggested in the report, "Promotion of STEM Education – Unleashing Potential in Innovation" (Education Bureau, 2016), integrated STEM education can be implemented with flexible use of school time and through timetabled lessons, and diverse activities outside schools, such as visiting science museums or reading STEM-related books.

Largely because of the "school-oriented" policy, it was stated in the consultation document of the Task Force on the Review of the School Curriculum (Education Bureau, 2019) that the pace and implementation strategy of STEM education varies greatly among schools, and the level of understanding and practice of STEM education in schools is very diverse. According to a statistical analysis by one local media outlet, over 60% of the schools included STEM education in conventional subjects and computer courses in their curriculum, and STEM activities are organised in a variety of forms, such as coding, 3D printing, or international STEM learning exchanges (LegCo Secretariat, 2020; Tang, 2019).

Wong et al. (2015) conducted a survey on the implementation of coding education with 42 primary and secondary schools (n = 42) in Hong Kong. Their results showed a growing trend in local schools promoting coding education, even though there were instructional challenges such as lack of teacher training and a unified curriculum. Kutnick et al. (2020) also surveyed students' opportunities for engineering learning in secondary schools. It was found that the students had different experiences of engineering learning, such as through Open Day activities, workshops organised in local universities, or International Chemistry Olympics. However, most students considered that there was an unavailability of engineering-related courses for them in school, due to the fact that their schools were not keen on engineering.

In a more recent project by So and Chiu (2020) which examined the school factors influencing students' STEM career aspirations, the results revealed that over 80% of the primary school students believed that STEM learning opportunities provided by their schools in Mathematics or Science were adequate, while 83.5% believed that their schools put much emphasis on STEM learning, and around 70% agreed that they had learnt something interesting from STEM activities/lessons and that the STEM activities/lessons were exciting. As for secondary schools, relatively fewer students (around 50%) agreed that they had sufficient STEM learning opportunities and support from schools for their STEM learning. A considerable percentage of secondary school students considered that the content taught in STEM lessons was interesting.

Out-of-school activities offered by different organisations

To provide a useful reference for students, parents, teachers, STEM providers, and education policymakers to make plans for STEM activities, the Croucher Foundation explored the out-of-school STEM ecosystem in Hong Kong. According to their reports (2017a, 2017b, 2019), it was found that the number of organisers and out-of-school STEM activities increased dramatically from 2015 to 2018. There were 144 organisers identified in 2015/2016, whereas in 2017/2018 the number increased to 471. The out-of-school STEM activities mapped in 2017/2018 numbered 3,065, which was nearly double the 2,074 activities mapped in 2015/2016. Most of the out-of-school STEM activities were science-related, and some covered one or more areas of STEM. The types were diverse, including but not limited to competitions, exhibitions, talks, workshops, courses, field trips, and camps.

These various STEM competitions, events, and courses have attracted the participation of many schools and students of different levels. Many of these STEM activities are free of charge, thus providing equitable learning opportunities for students from various backgrounds. The collaboration of organisers of the STEM competitions and events has become increasingly diversified, with contributions from universities, STEM industry sectors, and start-up companies.

For instance, the annual event, "Primary STEM Project Exhibition" (PSPE) co-organised by the Education University, EDB, and several associations, provides a platform for students from different schools to share the processes and products of their STEM inquiry through oral presentation and board exhibition/display (So, 2021). A research study has been conducted to analyse STEM activities demonstrated in students' projects in PSPE and found that more engineering and science activities than technology and mathematics activities were adopted by the primary students in their projects, and science activities significantly positively related to engineering and mathematics activities (So et al., 2018). The research evidence has implications for strategies to promote STEM integration.

The EDB has worked with The Hong Kong Academy for Gifted Education (HKAGE) to create challenging and out-of-school learning opportunities in

STEM domains, including STEM clubs (a student learning community piloted since 2019 to promote students' specialised and interdisciplinary STEM learning through project-based learning), the Hong Kong Youth Science & Technology Innovation Competition (including science fiction drawing and STEM activities for students), a summit regarding STEM policies and practices (for teachers, curriculum developers, leaders, and policymakers), and so forth. Moreover, there is also a talent scheme specifically designed for gifted primary or secondary students in STEM domains.

There are also other non-profit organisations that provide STEM learning resources, including the Hong Kong New Generation Cultural Association, the Hong Kong STEM Education Association, the STEM Education Centre, and so on. For example, the Hong Kong STEM Education Association is a non-profit organisation initiated by a group of experienced teachers and associations related to education. It provides STEM curricula for nearly a hundred primary and middle schools, organises STEM days or teacher workshops for schools, as well as STEM competitions (Hong Kong STEM Education Association, 2020). Instead of designing STEM curricula based on a project/goal/subject, the STEM curricula provided by the Hong Kong STEM Education Association are usually focused on certain technological processes/products such as Arduino and 3D printing.

Numerous companies, including many start-up companies, not only provide services for school STEM curricula but also informal STEM learning resources. For example, there are some activities organised by educational communities and industries to help promote students' entrepreneurship in STEM industries. For example, the Hong Kong Science and Technology Parks Corporation worked with several start-up companies and middle schools and universities to organise the first "STEM+E" scheme, in which students learned about the functions of the companies and engaged in problem solving in real contexts (Hong Kong Science and Technology Parks Corporation, 2020).

The Creative Education Unit of the Hong Kong Federation of Youth Groups was set up in 2009 and is now aiming to provide diversified free STEM education activities, such as robotics tournaments, creative coder competitions, and so on (Creative Education Unit, 2020). It built an open platform called a science makerspace, organising monthly activities for adolescents to engage in inquiry-based or design-based activities. In 2020, the makerspace organised many activities online. In addition, the Creative Unit also organised other blended STEM learning activities such as a "virtual innovation carnival", which includes lectures, hands-on experiments, and group discussions (Creative Education Unit, 2020). For example, the focus of the "virtual innovation carnival 2020" is a genetic engineering training program, which offers learning opportunities for knowledge and laboratory skills in molecular biology.

Some STEM education organisers focus their work on coding/computational thinking education. For example, the Hong Kong Jockey Club Charities Trust also provides STEM learning resources with joint effort from the City University of Hong Kong, Massachusetts Institute of Technology, and the

Education University of Hong Kong. Together they launched a program called "CoolThink@JC" which focuses on coding/computational thinking education specifically for primary level (Coolthink@JC, 2020). CoolThink@JC has not only supported coding education in many primary schools but also produced a series of free online workshops for students to learn coding and computational thinking with their parents. There are also many commercial companies that provide services regarding coding/computational thinking education. For example, a profitable platform for STEM learning named STEMhub is an online coding platform where students can learn coding online under the guidance of mentors and can engage in group projects (STEMhub, 2020).

However, it is worth noting that the boom in STEM learning resources is accompanied by the issue of utilitarianism. As educators have argued, in some STEM competitions or events that require high student ability, teachers of the students who participate as mentors tend to help to do the major tasks in these activities in order to bring rewards and honour to their schools (NgG & Tsang, 2020), which may hinder the development of students' ability building. The organisers of out-of-school STEM activities should thus carefully control the difficulty and age-appropriateness of competitions/events and regulate the over-interference of teachers, which may help release the full potential and motivation of students. Another issue would be that some out-of-school STEM activities appear to overemphasise technology products or processes, such as using Arduino, 3D printing, and coding, while neglecting the integration of other STEM disciplines. For example, learning coding or computational thinking alone without any scientific, engineering, or mathematical contexts may not meet the integrative nature of STEM. Moreover, the Covid-19 pandemic in 2020 makes it challenging for organisers to create meaningful platforms of online/blended STEM learning, discussion, and collaboration among learners and their parents.

University support for STEM education

Universities have an important role to play in preparing pre-service and in-service teachers with the necessary knowledge, evidence-based solutions, and innovations to support STEM education in schools. Universities contribute to STEM education by providing training to all teachers, or platforms for collaboration and community formation among schools, universities, and diverse STEM fields.

One project named "School-STEM Professionals' Collaboration" proposed a framework for enhancing collaborations between schools and STEM professionals for teacher professional development and student STEM learning (So et al., 2020). Teachers from several primary schools participated in STEM activity designs with support from educators and STEM experts from different fields, including environmental science, food science, information technology, or water and drainage engineering. Students in various classes of third to sixth grade engaged in these STEM activity designs to experience working as professionals in different STEM fields.

The results of So, et al. (2020) indicated that the "School-STEM Professionals' Collaboration" had a positive impact on teachers' conceptions of STEM education and STEM professionals. Teachers reported improved understanding of the STEM disciplines, student learning, and the contribution of STEM professionals. In addition, Chen et al. (2020) found that the "School-STEM Professionals' Collaboration" diversifying the stereotypes of STEM people could be advantageous to students as well. Students were found to show higher interest in STEM careers, but also more positive perceptions of STEM professionals. Particularly, girls were found to be more likely to improve their interests and alter their stereotypes with exposure to female role models, indicating the importance of considering gender-matching between students and STEM professionals while designing STEM activities.

Another project was conducted to support teachers' professional development related to STEM education for special needs students. The professional development programme aimed to provide special school teachers with opportunities to adapt STEM instruction to the needs of intellectually disabled (ID) students based on their knowledge of integrated STEM education and skills of scientific inquiry, technological literacy, engineering design, and mathematical thinking (So et al., 2019). Through analysing teachers' survey data, it was found that professional development enriched teachers' perceptions of STEM education with respect to its role in developing special needs students, and its advantages of cultivating students' interests in STEM learning.

In the follow-up research, teachers in special schools for children with intellectual disabilities were invited to work collaboratively with each other in their schools to plan and implement STEM learning (So et al., 2021). This research applied peer coaching as a supportive tool for facilitating the sharing of information and experiences among teachers (Hsieh et al., 2019) to improve their pedagogical knowledge and skills relevant to STEM education specific to students with intellectual disabilities. The positive perspectives collected from the teachers who participated in this research suggested that the technology use, the consideration of students' diversity, and the school involvement with peer coaching were important for STEM education in special schools.

Moreover, a project entitled "Joint university collaboration to develop students' competence and leadership in promoting integrated STEM education"[7] was implemented from 2017 to 2020. It provided useful insights into how collaborations among the universities in Hong Kong can be established to more effectively facilitate the development of STEM undergraduates and pre-service teachers responsible for STEM education. There were three main phases in this project, which allowed the university students to attend joint-university lectures/workshops outside their majors, and to experience learning projects through formulating joint-university, multi-disciplinary STEM learning communities, and to visit overseas STEM promoting, or STEM education centres to broaden their perspectives on STEM or STEM education.

A project (2019–2020)[8] funded by the Quality Education Fund was conducted for the purpose of enhancing STEM education in primary and secondary schools through integrating self-directed learning strategies. During the project, school leaders and teachers were provided with professional development workshops about self-directed learning, the use of Micro:bit software, and the assessment of STEM learning activities. It was expected that through the project, teachers would improve their capacities in curriculum-integrated STEM education, and students would consolidate their subject knowledge through STEM activities.

The perspectives of teachers on the keys to and limitations of STEM education

For schools to provide quality STEM education, the role of teachers is no doubt critical. It is important to listen to teachers' voices while moving STEM policies forward. The press release by the EDB (Cheng, 2018) listed suggestions by teachers for enhancing STEM education, including: (1) Teaching STEM in the school curricula rather than as a new curriculum; (2) Using student-centred approaches; (2) Formulating STEM teams through cross-subject collaborations; (4) Flexible use of existing resources provided by schools and the EDB; and (5) Enhancement of teacher professional development through school-based training.

Several studies (e.g., Croucher Foundation, 2017a, 2017b, 2019; Geng et al., 2019, Hong Kong Federation of Youth Group, 2018) have investigated Hong Kong teachers' perceptions and practices of STEM education. The Croucher Foundation (2017a) analysed the concerns of organisers of STEM activities outside schools and found that most were concerned about the lack of highly capable instructors, insufficient financial support, lack of equipment and venues, lack of support from schools, lack of parental support, and inadequate media coverage. The Croucher Foundation (2017a) also interviewed primary and secondary school teachers about their views on the limitations of in-school STEM activities. Some primary school teachers lacked confidence in giving clear instruction on the STEM concepts and theories due to not having a science background. For secondary school teachers, limited time to organise STEM activities was the main problem.

A project by Leung and Hu (2019) revealed that Hong Kong kindergarten teachers might lack the necessary STEM-related knowledge and pedagogical approaches for giving appropriate instruction and support for the children. Geng et al. (2019) found that almost half of Hong Kong teachers lacked self-efficacy in teaching STEM. The peak concerns among Hong Kong teachers were those related to "Information" (e.g., lack of resources), "Management" (e.g., lack of instructional time), and "Consequences" (e.g., impacts on students' learning). Teachers seek support in terms of teaching resources and funding, class and instructional implementation, and teacher training development, which are considered urgent needs from the perspective of the teachers' concerns. According to Yip and Chan (2019), it was noticed that even more experienced teachers needed help from professional consultants through building networks and partnerships within a broader STEM community.

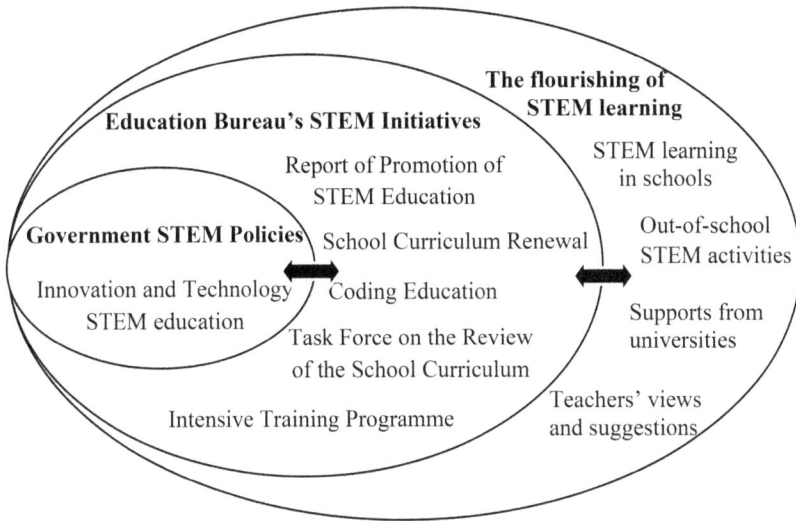

Figure 1.3 The interplay of STEM policies, initiatives, and implementation in Hong Kong.

Conclusion

Over the last two decades, STEM education has received and continues to receive increasing attention and priority in countries around the world, including in Hong Kong. This chapter has described how promoting STEM education has been historically integrated into policies in Hong Kong from 2015 to 2020. It is observed that innovation and technology and STEM education are mostly advocated as policy-driven initiatives in Hong Kong. The EDB has been stepping up efforts to promote STEM education, including the release of the report for STEM education, the ongoing renewal of the school curricula, introducing coding education in primary schools, and professional training for STEM teachers. Additionally, this chapter has discussed STEM activities in formal and informal contexts that address those policies, as well as the challenges or barriers faced by teachers in implementing interdisciplinary integration in existing school curricula.

It is encouraging to see that these policies and efforts are, to a large extent, helping to create a supportive policy context for STEM education in Hong Kong, contributing to the flourishing of STEM learning activities both in and outside of schools (see Figure 1.3). With a view to strengthening Hong Kong as a leading international I&T centre, it can be foreseen that concerted efforts will continue to be made to promote school-based and/or out-of-school STEM education, as nurturing STEM talent is one of the key elements of promoting I&T development. On teacher professional development, there will be ongoing work on organising more teacher trainings for core STEM teachers, in particular, to strengthen their pedagogical capacities in conducting quality STEM activities.

In considering how STEM education can be advanced to truly benefit students, the following are some suggestions based on the review of the implementation of government policies and the effort of the EDB. It can be observed that the presence of STEM integration in school curricula or out-of-school activities is still in its beginning stages due to the fact that science, mathematics, and technology are mostly taught as separate subjects, and engineering is inadequately addressed in the school curriculum. It is, therefore, important to put more emphasis on transparent and meaningful STEM connections across these school subjects. Particularly, mindful consideration in promoting coding education in the primary curriculum with careful deliberation of its role in STEM integration is warranted. For most STEM event organisers, their conceptualisation of the STEM activities as simply "coding" or "programming", while paying little attention to the integrated nature of STEM education is also problematic. These STEM learning opportunities appear to be oriented towards the technology strand, with a diminished focus on science and/or other disciplines. Hence, more emphasis should be placed on fostering multiple STEM skills other than just focusing on programming during STEM events. As for STEM teachers, it was found that a large percentage of them reported relatively low self-efficacy and expressed concerns regarding lack of resources or time, suggesting the need for ongoing support for teachers to be confident and competent in teaching STEM.

Overall, there is still a great deal that policymakers should do in their efforts to strengthen integrated STEM education in order to benefit students, families, and the society as a whole, as well as the building of modern knowledge economies globally.

Notes

1. https://www.edb.gov.hk/attachment/en/curriculum-development/renewal/CM/EDBCM20108E.pdf
2. https://tcs.edb.gov.hk/tcs/admin/courses/previewCourse/forPortal.htm?courseId=CDI020210873&lang=en
3. https://tcs.edb.gov.hk/tcs/admin/courses/previewCourse/forPortal.htm?courseId=CDI020210873&lang=en
4. https://stem.edb.hkedcity.net/en/home/
5. http://www.atec.edu.hk/stemcentre/index.html
6. https://www.edb.gov.hk/en/curriculum-development/4-key-tasks/it-for-interactive-learning/modular-computer-awareness-programme/index.html#9
7. https://www.ugc.edu.hk/doc/eng/ugc/activity/teach/triennium1619/EdUHK5.pdf
8. https://stemsdl21.eduhk.hk/

References

Chen, Y., Chow, S. C. F., & So, W. W. M. (2020). School-STEM professional collaboration to diversify stereotypes and increase interest in STEM careers among primary school students. *Asia Pacific Journal of Education*, 1–18.

Cheng, N. K. (2018). *STEM Education*. Retrieved from: https://www.edb.gov.hk/tc/about-edb/press/insiderperspective/insiderperspective20180402.html

Coolthink@JC. (2020). *What is computational thinking education*. Retrieved November 25, 2020, from https://www.coolthink.hk/coolthink-classroom/

Creative Education Unit. (2020). *Nurturing of Local Talent*. Retrieved November 19, 2020, from https://ce.hkfyg.org.hk/

Dutta, S., Lanvin, B., & Wunsch-Vincent, S. (2020). *Global Innovation Index Report 2020: Who Will Finance Innovation?* Retrieved from: https://www.wipo.int/edocs/pubdocs/en/wipo_pub_gii_2020.pdf

Education Bureau. (2014). *The Fourth Strategy on Information Technology in Education: Realising IT Potential, Unleashing Learning Power, a Holistic Approach*. Retrieved from: https://www.edb.gov.hk/attachment/en/edu-system/primary-secondary/applicable-to-primary-secondary/it-in-edu/ITE4_report_ENG.pdf

Education Bureau. (2016). *Report on Promotion of STEM Education: Unleashing Potential in Innovation*. Retrieved from: https://www.edb.gov.hk/attachment/en/curriculum-development/renewal/STEM%20Education%20Report_Eng.pdf

Education Bureau. (2019). *Task Force on Review of School Curriculum: Consultation Document*. Retrieved from: https://www.edb.gov.hk/attachment/en/curriculum-development/renewal/taskforce_cur/TF_CurriculumReview_Consultation_e.pdf

Education Bureau. (2020). *Optimise the Curriculum for the Future: Foster Whole-person Development and Diverse Talent. Task Force on Review of School Curriculum Final Report*. Retrieved from: https://www.edb.gov.hk/attachment/en/curriculum-development/renewal/taskforce_cur/TF_CurriculumReview_FinalReport_e.pdf

Duncan, C., Bell, T., & Tanimoto, S. (2014, November). Should your 8-year-old learn coding? In *Proceedings of the 9th Workshop in Primary and Secondary Computing Education* (pp. 60–69). New York, United States, Association for Computing Machinery.

Geng, J., Jong, M. S. Y., & Chai, C. S. (2019). Hong Kong teachers' self-efficacy and concerns about STEM education. *The Asia-Pacific Education Researcher*, *28*(1), 35–45.

Hong Kong Federation of Youth Group. (2018). *STEM Education in Secondary Schools: Improving Resource Utilization*. Retrieved from: https://yrc.hkfyg.org.hk/en/2018/01/14/stem-education-in-secondary-schools-improving-resource-utilization-2/.

Hong Kong Science and Technology Parks Corporation. (2020). Press room. *HKSTP*. Retrieved from: https://www.hkstp.org/zh-hk/press-room/

Hong Kong STEM Education Association. (2020). Hong Kong STEM Education. *Hkstemedu*. Retrieved from: https://www.hkstemeducation.com/

Hsieh, F. P., Lin, H. S., Liu, S. C., & Tsai, C. Y. (2019). Effect of peer coaching on teachers' practice and their students' scientific competencies. *Research in Science Education*, *49*, 1–24. doi: https://doi.org/10.1007/s11165-019-9839-7.

Kutnick, P., Lee, B. P. Y., Chan, R. Y. Y., & Chan, C. K. Y. (2020). Students' engineering experience and aspirations within STEM education in Hong Kong secondary schools. *International Journal of Educational Research*, *103*, 101610.

LegCo Secretariat. (2020). *Nurturing of Local Talent*. Retrieved from: www.legco.gov.hk/research-publications/english/1920rb03-nurturing-of-local-talent-20200601-e.pdf

Leung, W. M., & Hu, X. A. (2019). "We want STEM": Exploring digital toys in a Hong Kong kindergarten. *Journal of Education and Human Development*, *8*(4), 82–93.

Lye, S. Y., & Koh, J. H. L. (2014). Review on teaching and learning of computational thinking through programming: What is next for K-12? *Computers in Human Behavior*, *41*, 51–61.

NgG, C. K. W., & Tsang, H. K. (2020). Diverse strategies of STEM education in Hong Kong schools: The importance and contributions of design and technology teachers. *The Hong Kong Teachers' Centre Journal* (in Chinese), *18*, 45–55. Retrieved from: https://www.edb.org.hk/irooms/eservices/T-surf/Content/Documents/HKTC%20Volume%2018.pdf

Selby, C., & Woollard, J. (2013). *Computational Thinking: The Developing Definition*. University of Southampton Institutional Research Repository. Retrieved from: http://eprints.soton.ac.uk/356481

So, H. J., Jong, M. S. Y., & Liu, C. C. (2020). Computational thinking education in the Asian Pacific region. *The Asia-Pacific Education Researcher, 29*, 1–8.

So, W. M. W., Zhan Y. Chow, C. F., & Leung, C. F. (2018). Analysis of STEM activities in primary students' Science projects in an informal learning environment. *International Journal of Science and Mathematics Education, 16*(6), 1003–1023.

So, W. M. W. & Chiu, W. K. S. (2020). Challenges and Opportunities with Hong Kong Students' Science, Technology, Engineering and Mathematics Aspirations. *Public Policy Research Funding Scheme*. Retrieved from: https://www.pico.gov.hk/doc/en/research_report(PDF)/2018.A5.041.18C_Final%20Report_Prof%20So.pdf

So, W. M. W., He, Q., Chen, Y., & Chow, C. F. (2020). School-STEM professionals' collaboration: A case study on teachers' conceptions. *Asia-Pacific Journal of Teacher Education*, 1–19. Retrieved from: https://www.tandfonline.com/doi/full/10.1080/1359866X.2020.1774743

So, W. M. W., Li, J., & He, Q. (2019). Teacher professional development for STEM education: Adaptations for students with intellectual disabilities. *Asia-Pacific STEM Teaching Practices* (pp. 83–102). Singapore: Springer.

So, W. M. W., He, Q. W., Cheng, I. N. Y., Lee, T. T. H., & Li, W. C. (In press). Teachers' Professional Development with Peer Coaching to Support Students with Intellectual Disabilities in STEM Learning. *Educational Technology & Society*.

So, W. M. W. (2021). Does Computation Technology Matter in Science, Technology, Engineering and Mathematics (STEM) Projects? *Research in Science & Technological Education*, 1-19.

STEMhub. (2020). About STEMhub. *Stemhub*. Retrieved from https://www.stemhub.com/

Tang, W. (2019). Programming and STEM Should be Included in School Curriculum. *Ejinsight*. Retrieved from: https://www.ejinsight.com/eji/article/id/2099537/20190401-programming-and-stem-should-be-included-in-school-curriculum

The Croucher Foundation. (2017a). *The Out-of-School STEM Ecosystem in Hong Kong: An Exploratory and Investigative Study 2015/2016*. Retrieved from: https://croucher.org.hk/wp-content/uploads/2017/02/CF_STEM_study2015-16.pdf

The Croucher Foundation. (2017b). *The Out-of-School STEM Ecosystem in Hong Kong Second Report 2016–2017*. Retrieved from: https://croucher.org.hk/wp-content/uploads/2017/02/CF_STEM_study2016-17-1.pdf

The Croucher Foundation. (2019). *The Out-of-School STEM Ecosystem in Hong Kong Third Report 2017–2018*. Retrieved from: https://croucher.org.hk/wp-content/uploads/2019/08/CF_2019_0719.pdf

The Curriculum Development Council. (2017a). *Science Education: Key Learning Area Curriculum Guide (Primary 1 – Secondary 6)*. Retrieved from: https://www.edb.gov.hk/attachment/en/curriculum-development/renewal/SE/SE_KLACG_P1-S6_Eng_2017.pdf

The Curriculum Development Council. (2017b). *Technology Education: Key Learning Area Curriculum Guide (Primary 1 – Secondary 6)*. Retrieved from: https://www.edb.gov.hk/attachment/en/curriculum-development/renewal/TE/TE_KLACG_P1-S6_Eng_2017.pdf

The Curriculum Development Council. (2017c). *Mathematics Education: Curriculum Guide (Primary 1 – Secondary 6)*. Retrieved from: https://www.edb.gov.hk/attachment/en/curriculum-development/renewal/ME/ME_KLACG_P1-S6_Eng_2017.pdf

The Curriculum Development Council. (2017d). *General Studies: Key Learning Area Curriculum Guide (Primary 1–6) (2017)*. Retrieved from: https://www.edb.gov.hk/attachment/en/curriculum-development/renewal/GS/GS_KLACG_P1-6_Eng_2017.pdf

The Curriculum Development Council. (2020). *Computational Thinking – Coding Education: Supplement to the Primary Curriculum*. Retrieved from: https://www.edb.gov.hk/attachment/en/curriculum-development/renewal/CT/CT%20Supplement%20Eng%20_2020.pdf

The Government of the Hong Kong Special Administrative Region. (2015). *The Executive's 2015 Policy Address: Uphold the Rule of Law Seize the Opportunities Make the Right Choices*. Retrieved from: https://www.policyaddress.gov.hk/2015/eng/pdf/PA2015.pdf

The Government of the Hong Kong Special Administrative Region. (2016). *The Executive's 2016 Policy Address: Innovate for the Economy Improve Livelihood Foster Harmony Share Prosperity*. Retrieved from: https://www.policyaddress.gov.hk/2016/eng/pdf/PA2016.pdf

The Government of the Hong Kong Special Administrative Region. (2017). *The Chief Executive's 2017 Policy Address: We Connect for Hope and Happiness*. Retrieved from: https://www.policyaddress.gov.hk/2017/eng/pdf/PA2017.pdf

The Government of the Hong Kong Special Administrative Region. (2018). *The Chief Executive's 2018 Policy Address: Striving Ahead Rekindling Hope*. Retrieved from: https://www.policyaddress.gov.hk/2018/eng/pdf/PA2018.pdf

The Government of the Hong Kong Special Administrative Region. (2019a). *The Chief Executive's 2019 Policy Address Supplement: Treasure Hong Kong: Our Home*. Retrieved from: https://www.policyaddress.gov.hk/2019/eng/pdf/supplement_full.pdf

The Government of the Hong Kong Special Administrative Region. (2019b). *Legislative Council Panel on Education 2019 Policy Address Education Bureau's Policy Initiatives*. Retrieved from: https://www.edb.gov.hk/en/about-edb/legco/policy-address/2019_Panel_on_Education_Eng.pdf

The Government of the Hong Kong Special Administrative Region. (2020a). *The Chief Executive's 2020 Policy Address Striving Ahead with Renewed Perseverance*. Retrieved from: https://www.policyaddress.gov.hk/2020/eng/pdf/PA2020.pdf

The Government of the Hong Kong Special Administrative Region. (2020b). *Legislative Council Panel on Education 2020 Policy Address Education Bureau's Policy Initiatives*. Retrieved from: https://www.legco.gov.hk/yr20-21/english/panels/ed/papers/ed20201204cb4-231-1-e.pdf

Weintrop, D., Beheshti, E., Horn, M., Orton, K., Jona, K., Trouille, L., & Wilensky, U. (2016). Defining computational thinking for mathematics and science classrooms. *Journal of Science Education and Technology, 25*(1), 127–147.

Wong, G. K., Cheung, H. Y., Ching, E. C., & Huen, J. M. (2015). School perceptions of coding education in K-12: A large scale quantitative study to inform innovative practices. *Paper presented at the 2015 IEEE International Conference on Teaching, Assessment, and Learning for Engineering (TALE).* IEEE.

Yip, V. W., & Chan, K. K. H. (2019). Teachers' conceptions about STEM and their practical knowledge for STEM teaching in Hong Kong. In Hsu, Y. S., & Yeh, Y. F. (Eds.) *Asia-Pacific STEM Teaching Practices: From Frameworks to Practices* (pp. 67–81). Singapore: Springer.

2 STEM education in Malaysia

Policies to implementation

*Lilia Halim, Lay Ah Nam, and
Edy Hafizan Mohd Shahali*

Introduction

Malaysia has always heavily stressed providing Science, Technology, Engineering, and Mathematics (STEM) education at all levels of education. STEM education is a utilitarian tool for Malaysia as a developing nation that strives for an economy that is knowledge and innovation-based. The workforce in the field of STEM has dominated the needs of the domestic labour market in Malaysia. In 2017, Malaysia reported that 7% of the country's gross domestic product (GDP) is from STEM-related industries, and the percentage is projected to rise to 45% by 2021 (MOSTI, 2018). However, amidst rapid growth in STEM-related industries, the number of students taking STEM-related subjects has decreased yearly (MOE, 2018a). The number of STEM students was low and far from reaching the national target. Despite the demand for STEM-related occupations, which were the top emerging jobs, the STEM skilled workforce was on the decline. The Malaysian Ministry of Education (MOE) reported that in 2018, in upper secondary, only 83,608 (22%) of Malaysian students were in the STEM stream as compared to 92,956 (24%) in 2017, with a reduction of 9,348 students (MOE, 2018a).

The decrease in student participation in STEM-related subjects and courses was occurring at all levels of education in Malaysia. At the post-secondary level, only 5,865 students took Science subjects in the 2017 *Sijil Tinggi Persekolahan Malaysia* (STPM) – similar to GCE A-Level – compared to 4,773 in 2018 (MOE, 2018b). In 2017, the number of students studying in the field of STEM at the tertiary level was 211,435. (MOE, 2018b). As a comparison, this was much lower than the 327,120 students who enrolled in social sciences in the same year.

Many factors contributed to influencing student participation and the quality of student learning in the STEM fields, namely (1) Limited awareness about the value of STEM learning and career among students and parents, (2) Perceived difficulty of STEM subjects, (3) Content-loaded curriculum with less emphasis on relevance to everyday life and opportunities to be creative and innovative, (4) Teaching and learning approaches are teacher-centred, and (5) Limited and outdated infrastructure (MOE, 2013, pp. 4–6).

DOI: 10.4324/9781003099888-2

Policies related to STEM education

In recognition of STEM in ensuring Malaysia's economic competition; STEM education has been the country's focus for the past decade. Thus, building and strengthening human capacity in the competencies of STEM is paramount. The Malaysian government has systemically incorporated STEM education through several unique STEM policies in Malaysia. The main STEM education-related policies are:

60:40 Science/Technical: Arts policy

The 60:40 Science/Technical: Arts (60:40) Policy in education, established in 1967 by the Malaysia Higher Education Planning Committee, was enforced in 1970. It targets 60% among all upper high school students concentrating on Science while 40% participating in arts. The MOE delineates strategies and initiatives to achieve the 60:40 policy. In public schools, to qualify as a student in the Science stream, the student has to take at least two pure science courses (such as Chemistry and Biology) at the upper secondary level. Such a combination of subjects would entitle the student to register for STEM undergraduate courses. The government established specialised STEM schools as part of the 60:40 policy. The 60:40 policy is also the foundation of the 2020 Human Resources Roadmap by the Ministry of Science, Technology, and Innovation (MOSTI), thus endorsing that the achievement of the 60:40 goal has become more significant and critical than before. MOSTI has fine-tuned its 60:40 strategy in acknowledgment of the increasing economic value of technical and vocational training. Currently, 60% of secondary education refers to either Science or Arts streams, while the other 40% relates to Technical and Vocational career pathways.

Malaysia education blueprint (MEB) 2013–2025

The MEB 2013–2025 is the National Education Blueprint developed from a systematic analysis of its educational system. The government embarked on the review to better prepare children for future challenges and address the public's demand for a quality education system. For 15 months, the Ministry relied on inputs from stakeholders from all over the country. The outcome was a Malaysian Education Blueprint that provides Malaysia's vision of a first-class education within the Malaysian socio-culture context, particularly the nature of students' outcomes resulting from the educational system. The MEB recommended 11 institutional and operational improvements needed to accomplish the goals, which are on par with the global standards.

One of the strategic thrusts in MEB is to increase the quality of STEM education. A National STEM Action Plan was established by the MOE, MOSTI, and Ministry of Higher Education (MOHE), involving all federal agencies and partnerships with the private sector. This action plan addressed many facets of STEM, such as awareness, recruitment, infrastructure, research, and job opportunities.

MOE also introduced a programme known as the *STEM Education Initiative* as part of the National STEM Action Plan. The aim of the STEM education initiative was two-fold. Firstly, it was to prepare students with the STEM competencies to address the challenges posed by STEM advancement. Secondly, was to ensure that Malaysia has the number of STEM graduates to fulfill the work sectors that support its STEM-based economy (MOE, 2013). Three core strategies under the STEM education initiative revolved around (1) Raising students' interest in STEM education, (2) Upskilling teachers' competencies, and (3) Enhancing parents' and students' awareness toward STEM.

Strategy one: Raise students' interest

Raising students' interest in STEM involves both introducing new subject matters and active learning and teaching approaches. The new subject matters are relevant to promote higher-order thinking skills (HOTS) – thinking skills needed in STEM inquiry activities. The active learning and teaching approaches set in real-world problems are necessary for the STEM learning environment.

ENHANCED LEARNING APPROACHES THROUGH
DEVELOPED STEM MATERIAL

The MOE initiated teaching and instructional resources for STEM education (MOE, 2018b). Examples of learning materials developed are as follows.

- The teaching and learning resource materials known as the Science, Technology, Engineering, and Mathematical Resource Material (BSTEM) Series, which employed a project-based approach as guidelines for educators to implement STEM in the classroom.
- A compilation of immersive videos known as the Blended Learning Open Source Science or Math Studies (BLOSSOMS) developed to support teaching and learning practices. In 2015, a partnership with the Malaysia University of Technology and the Massachusetts Institute of Technology (MIT) supported the initiative.
- Virtual STEM courses known as Massive Open Online Course (MOOCs) developed by Universiti Putra Malaysia (UPM), in collaboration with MOE. The course goals were to improve teachers' skills and knowledge in the STEM field, encourage teachers to incorporate STEM within their respective subjects and provide teachers with an easy access forum that enabled them to learn independently, regardless of time and place, through accessible learning modules.
- Selected STEM expert teachers assisted in the development of the STEM Teaching and Learning Multimedia Module. The module's purpose was to provide training for STEM teachers in diversifying STEM teaching and learning pedagogy with computer and information technology applications.

- In collaboration with MOE, public universities developed STEM comics aimed at primary school students to attract their interest in STEM. The STEM comics drew on a simple storytelling approach with the application of HOTS aligned with the elementary school syllabus.

Based on the study of Talent for Survival (Deloitte Touche Tohmatsu Limited, 2016), STEM skills and knowledge are important, but a workforce with a balance of technical and soft skills is more beneficial to the country in exploring the digital world's era in a knowledge-based economy. For them to succeed internationally, students must acquire 21st-century skills. In the MEB 2013–2025, the goal of promoting thinking skills was to empower every student with experience, reasoning skills, leadership ability, linguistic skills, morals and religion, and nationhood. The MOE has outlined strategies to enhance the national primary and secondary STEM curriculum to achieve this goal.

Enculturation of thinking skills: HOTS have been the focus of educational reforms in recent years and are more significant now due to the demands of the Fourth Industrial Revolution (4IR). The STEM curriculum in Malaysia provides opportunities for developing thinking skills in addition to developing students with soft skills. The STEM curriculum also aims to develop students that are resilient, team players, curious, and patriotic (BPK, 2014). These qualities are essential for students to be able to compete at national and global levels. Thus, the content and learning standards, including assessment, are aligned to develop HOTS. The incorporation of HOTS through teaching and learning activities utilises thinking strategies and reasoning skills to promote thinking skills (BPK, 2014). Students are expected to apply thinking strategies and reasoning skills in problem solving and decision-making, innovating, and creating (BPK, 2014).

Develop computational and technological literacy: The STEM curriculum also aims to prepare students to master computational and technology literacy, including computational thinking (CT) – a literacy that is no more exclusive for computer scientists but a necessary basic literacy in meeting the challenges in the era of rapid technological development, namely the Industrial Revolution 4.0. In general, CT refers to inspecting problems methodically then formulating solutions based on concepts in computer science. CT helps students organise, analyse, and present data or ideas logically and systematically to efficiently solve complex problems. Students must learn to use computers and computational technologies to solve problems as the technology and computing processes drive the nature of work in the 4IR environment. For instance, analysis of large data sets employs computational tools, and the act of processing large data sets will be both familiar and essential in the digital environment. As the volume of data increases exponentially, computational tools will be required to make sense of the information (Davies et al., 2011).

The STEM curriculum has been also enhanced and aims to develop students' technological literacy as students in the 21st century need to act as knowledgeable technology users as well as technology creators and innovators. In Malaysia, Design and Technology (D&T) is a compulsory subject for primary level II (Year 4–6) and lower secondary level (Form 1–3). Students learn design skills using technology by integrating various disciplines and skills, namely technical, agricultural technology, and household science. The curriculum emphasises basic knowledge and design skills, basic carpentry, and technology to produce quality products. In line with the 21st century and the 4IR encounters, a review of the D&T curriculum in 2020 emphasised the production process by integrating design and technology skills and programming. The enhanced curriculum focuses on design appreciation, design process, technical knowledge and technology application, product manufacturing, and product design evaluation. Students apply knowledge and skills through design activities and produce meaningful products.

In 2017, a new subject, Basic Computer Science (BCS) was introduced to students of the lower secondary level. The BCS curriculum focuses on producing students with CT skills who can unravel problems, design systems, and comprehend human behaviour – an introductory computer science principle. Computer Science (CS) offered at the upper secondary level is a continuation of the BCS at the lower secondary level. The CS subject aims to develop students' computational and critical thinking skills through designing, developing, and creating new applications.

Integrate science and technology elements into non-STEM subjects: One would have to draw upon STEM and non-STEM subjects' content, thinking, and disposition in today's complex world. Thus, integrating STEM elements such as (1) Related knowledge and concepts, (2) Scientific thinking skills and processes, (3) Scientific attitudes, and (4) The use of technology is aimed at enhancing the students' learning experiences. Furthermore, the application of Information and Communication Technology (ICT) in teaching and learning can promote students' interest in science and technology, encourage students to be innovative, make teaching and learning more relevant, and enhance students' ICT literacy. Under the MEB 2013–2025, the number of ICT devices will gradually increase until the student-to-device ratio reaches 10:1. Each student can use and leverage the devices for effective learning. Basic competencies and literacy in ICT are the basis for propelling innovations in pedagogy. Thus, the MOE is also aiming to have all teachers meet a minimum level of ICT literacy.

Encourage transdisciplinary activity and entrepreneurial innovation: Science, technology, and innovation (STI) are the drivers of Malaysia's socio-economic change towards a high-income, inclusive, and sustainable developed country. Exposure to research, development, design, commercialisation, and entrepreneurship (R, D, I, C, & E) activities at the school level aim to increase students' interest in R, D, I, C, & E activities. Strategies aimed at promoting R, D, I, C, & E activities in educational institutions include:

- Expanding academic programmes that contain R, D, & I in the early stages of schooling to provide awareness of R, D, & I's potential to generate wealth through commercialisation.

- Assisting students in registering their innovations as intellectual property for commercial purposes. The R, D, & I activities conducted in silos without commercialisation and entrepreneurship elements are among Malaysia's challenges (MOSTI, 2018). Valuation and commercialisation of intellectual property enable the promotion of R, D, & I among students.
- Expanding the entrepreneurship programme to intensify STEM students' participation in R, D, I, C, & E. STEM students should also learn entrepreneurial skills. Entrepreneurship leverages and applies innovation to introduce new processes, services, and products to markets and communities.

Realising that discovery and innovation are usually catalysed by integrating knowledge and practice in STEM and non-STEM, interdisciplinary activities and innovations are encouraged. Students are encouraged to integrate engineering practices, science content and practices, and mathematical content and practices (including other disciplines) to create technologies to solve real-world problems through collaborative teamwork.

The *Bitara-STEM: Science of Smart Communities Programme* (*Bitara*-STEM) is an example of an interdisciplinary programme conducted by the Faculty of Education, National University of Malaysia (UKM) for middle secondary school students. The programme encouraged students' innovation through a hands-on project that is interdisciplinary in nature in an outreach learning environment by drawing upon the five phases of an engineering design process (Mohd Shahali et al., 2019).

The programme consists of four different modules: Energy, Urban Infrastructure, Transportation, and Wireless Communication (Figure 2.1). Each module entails STEM content from Earth Science, Physical Science, and Life Science, with daily themes overlapping these three content areas. These topics reflect the components involved in a model of smart cities. Planning for a smart city entails students playing the role of different STEM professionals (especially engineers and scientists) in numerous arenas (Mohd Shahali et al., 2017). The programme is mainly to ignite students' interest in STEM, increase students'

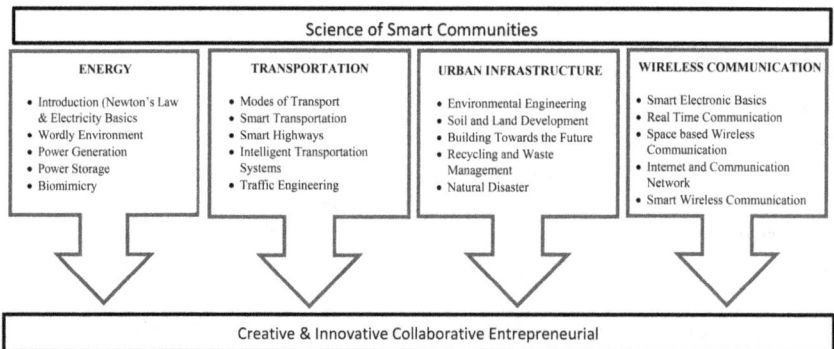

Figure 2.1 Bitara-STEM: Science of Smart Communities Programme.

21st-century skills (such as collaboration, entrepreneurship, innovation, research skill, HOTS, etc.) and improve students' creativity to solve problems. *Bitara-STEM* approach draws on real settings driven by project-oriented problem-based learning (PoPBL).

Strategy two: Development of educators' competencies

In any educational innovation, the success of its implementation would depend on the educators' competencies. The MEB has outlined strategic initiatives to ensure the competencies development of the educators.

ESTABLISHING THE NATIONAL STEM CENTRE

The National STEM Centre (Pusat STEM Negara), established in 2018, provides meaningful learning experiences and ensures teachers design lessons based on new methods and approaches. The centre is responsible for planning, implementing, and coordinating STEM-related activities, especially continuous professional development for STEM teachers. Among the National STEM Centre's key aspects for teacher empowerment is improving teacher pedagogical knowledge and skills in integrated STEM teaching and learning. To familiarise the teachers with integrated STEM pedagogy (i.e., problem-based learning and project-based learning), MOE had trained Master Trainers at the national, state, and school levels to prepare teachers for this learning method. Realising that cascading training has its disadvantages (Bush et al., 2019), the School Improvement Specialists Coaches (SISC+) programme was established to address the limitations while meeting more contextual training needs. The coaches selected are experienced teachers with outstanding credentials in pedagogy. Thus, it is the coach's role to enhance the practicing school teachers' pedagogies. The coaches work at the district levels; therefore, they can provide in-service training that meets the stakeholder's requests. Among the types of in-service training are the introduction and implementation of active learning-based pedagogies using appropriate technology to raise students' interest and motivation in STEM in the classroom. Such in-service training that is contextual and targeted would likely increase the success rate in integrating STEM in their classroom lessons. The coaches themselves also undergo professional development as part of their in-service training.

The National STEM Centre has also implemented various programmes involving the development of STEM teacher professionalism and enhancing school leaders' and laboratory assistants/technicians' STEM competencies and confidence. Among the programmes involving teacher professionalism is introducing Inquiry-Based Science Education (IBSE) approaches to science teachers. Through these approaches, teachers act as intermediaries of knowledge and facilitators. Students are actively involved in mastering STEM subjects through the collection and synthesis of information and inquiry skills. The student-centred approaches promote curiosity and produce students who can think critically, creatively, and innovatively. The training module draws on the IBSE module developed by the *La main à la pâte* Foundation, France. The IBSE workshop's implementation involved two phases: Training of the National Master Trainer (TPU) at the national level and training of the Master Trainers (JU IBSE) at the state level. Each master trainer undergoes 60 hours of training to qualify as a state JU IBSE. There are workshops organised by the National STEM Centre like Workshop on Nourishing Creativity through Transdisciplinary Mathematical Education in collaboration with the Southeast Asian Ministers of Education Organisation (SEAMEO), Regional Centre for Education in Science and Mathematics (SEAMEO RECSAM), Climate Change Education Training Workshop for IBSE Master Trainers in partnership with Academy of Sciences Malaysia (ASM), and International Science, Technology and Innovation Centre (ISTIC) and many more.

DEVELOP STEM TEACHERS' COMPETENCY FRAMEWORK

The STEM Teachers' Competency Framework (Table 2.1) developed by the Teacher Education Division (BPG, In press), MOE, aims to assess teachers' teaching quality in STEM. The framework focuses on the critical competencies needed to produce quality STEM teachers. Through this framework, teachers'

Table 2.1 Definition of each competency

Competency	Definition
Beliefs about STEM learning	Teachers are to possess, express, and share beliefs about the importance and necessity of STEM and future careers.
STEM content knowledge and skills	Teachers understand and deliver STEM subject content knowledge, processes, and skills according to the relevant STEM subject or field.
STEM-related pedagogies	• Teachers are to master teaching and learning strategies in the classroom that contribute to the optimal development of global competitiveness competencies. • Teachers are to plan and deliver teaching and learning strategies based on students' capabilities and inclinations.
Real-world application	• Teachers can identify and embed STEM subject content in daily, real-life scenarios. • Teachers are to design real-world experience on STEM subject content.
Capacity to integrate STEM and non-STEM	• Teachers understand the content knowledge, processes, and skills across STEM and non-STEM in teaching and learning. • Teachers exhibit collaborative skills to work with STEM or non-STEM teachers in an integrated manner across classes. • Teachers can identify and deliver skills relevant to STEM careers across STEM and non-STEM disciplines.
Data, digital, and technological literacy	Teachers understand and apply concepts, methods, and tools for data, digital, and technology usage safely and ethically.

quality can be monitored and evaluated to develop and plan for relevant teacher development programmes from time to time.

STEM LEADERSHIP

Apart from programmes that target teachers, in-service programmes for school administrators also provide awareness of STEM education to school administrators. The *Aminuddin Baki Institute* (IAB), an institute responsible for school leaders' professional development, also conducts a STEM Leadership Course. The STEM Leadership Course aims to help school leaders (headmasters and principals) increase their awareness, knowledge, skills, and leadership in STEM education through curriculum and co-curricular activities. The course's specific objectives were to raise awareness of the importance of STEM education, improve the understanding of STEM education policies and concepts, and improve skills in integrating STEM elements in teaching and learning. School leaders are encouraged to develop the School's Transformation Intervention Plan to achieve the target of strengthening STEM education. The co-curricular activities and actions plan is part of the transformation plan. In 2018, a total of eight STEM Leadership courses were offered by IAB with the involvement of 240 school principals. The implementation of the in-service courses showed a positive impact on strengthening STEM education in the schools involved (MOE, 2018b).

Strategy three: Building public and students' awareness

In addition to enhancing STEM education through the formal education system, the informal learning environment can complement STEM education in the formal system. As a result, there has been an increase in STEM outreach programmes as a means to promote students' and public awareness of STEM and STEM education.

ENABLE QUADRUPLE HELIX NETWORK AND LEVERAGE
CORPORATE SOCIAL RESPONSIBILITY

The MOE is taking the initiative to fully engage the public, namely the parents, and the private sectors in developing the STEM learning ecosystem in schools and outside of schools. Benefits gained through this engagement will indirectly have a comprehensive impact on students' learning. For example, involving the public (such as industries and non-profit and community organisations) can bring in resources (in the form of monetary or in-kind capabilities) which can complement the shortages of what the public education sector may be able to provide.

In ensuring the sustainability of the STEM learning ecosystem, strategic cooperation with various parties needs to be strengthened by building partnerships to include expertise sharing at the national and international levels. This strategic collaboration allows stakeholders to collaborate with multiple parties who have advantages in efficiency, knowledge, and resources in the STEM field. Several strategies to promote strategic cooperation include: (1) Enabling quadruple helix (government-academia-industry-community) network collaboration in STEM programmes at the national and international levels and (2) Leveraging corporate social responsibility. For example, the National STEM Centre has collaborated with various parties to organise professional development programmes for teachers and also STEM programmes for students. The programmes can be found on the website of National STEM Centre and in the MEB 2018 annual report (https://www.padu.edu.my/annual_report/2018-2/).

Among them are:

- IBSE programme with ASM and ExxonMobil Malaysia. A total of 1200 teachers gained a deeper understanding of the inquiry-based approach.
- Inquiry-Based Mathematics Education (IBME) programme with the Tun Hussein Onn Teachers Foundation. Several mathematics learning modules have been developed based on the inquiry approach. The modules are used in mathematics teacher continuous development programmes.
- Programme Duta Guru with Yayasan PETRONAS to strengthen STEM teachers' capability and continuously improve their quality of teaching. This programme aimed at upskilling 4,500 STEM teachers serving as role models to enhance HOTS through STEM education, as a foundation to improve Malaysia's competitiveness. The STEM Hub is, amongst several others, an infrastructure resource in the programme managed by Petrosains (a Malaysian science and technology museum) to support teacher upskilling by connecting teachers to their peers for collaboration (https://www.yayasanpetronas.com.my/program/dutaguru/).
- "1,2,3…Code!" programme with ISTIC and Universiti Teknologi Malaysia (UTM). Three experts from the Foundation *La main à la pâte* and six computer scientists from UTM and ISTIC were involved (https://www.fondation-lamap.org/fr/123codez).
- The IET Faraday Challenge with the Institute of Engineering and Technology (IET) Malaysia Network, Malaysia Digital Economy Corporation (MDEC), and Honeywell Malaysia. The challenges gave students an opportunity to draw upon their knowledge of design and engineering to produce prototype that solves real-life, genuine engineering problems. Experts from IET Malaysia were involved in facilitating the participants.

MOE also cooperates with various local agencies to increase students' interest in STEM. To achieve this goal, MOE worked with other government agencies (and private sectors) to organise various STEM competitions, hands-on STEM sessions, and STEM camps. Among the programmes was School Lab Competition, co-organised by MOE and British Council Malaysia, in collaboration with Malaysian Industry-Government Group for High Technology (MIGHT) and Petrosains. In this competition, students used their creativity to create a video (in the form of an act, a song, and others) describing a STEM concept and uploaded it to YouTube for evaluation at the state and eventually at the national level. The competition created a lot of interest among students and teachers. Students expressed that their interest in STEM had increased tremendously through this competition, even though their videos did not make it for the final event at the state or national level (Shahali et al., 2017).

MOE has also managed to attract and engage volunteers from various STEM-related organisations to raise awareness and cultivate students' interest in pursuing STEM careers. As a result, the National STEM Association, established in 2017, comprises government agencies, universities, private organisations, and industry. The primary role of National STEM Movement was to promote and implement the Enhancing STEM Education Initiative. Some of the programmes conducted were STEM mentoring, STEM seminars, and STEM carnivals. The target participants were usually teachers, students, and parents.

A particular STEM programme that is worth mentioning is the *Bitara*-STEM outreach programme. This outreach programme resulted from collaborative work between the New York Academy of Sciences (NYAS) and New York Polytechnic School of Engineering (NYU-Poly) with STEM-related faculties from UKM. To date, the programme has reached more than 3000 students and teachers. Since 2015, UKM, in collaboration with ExxonMobil Malaysia, extended the *Bitara*-STEM Programme's implementation for teachers and students, involving academia from UKM and engineers/scientists from ExxonMobil.

In addition, there is series of organised colloquiums and conferences on STEM aimed at creating a network of partnerships between policymakers, academicians, educators, counsellors, government and non-government agencies, and sharing best practices for successful STEM education. The goal of colloquiums and conferences is to create awareness about STEM and STEM-related careers among school administrators, teachers who teach STEM subjects, and school counsellors. Besides, it also provides academia and other researchers opportunities to share their research findings and creative ideas in implementing innovative teaching and learning processes related to STEM and creates a platform for professional networking among academics and educators. To date, there have been many series of colloquiums and conferences organised under the STEM Education initiative. For example, the STEM Education Colloquium series and The International Conference on STEM Education (ICSTEM) have been held since 2017.

Collaboration between schools and parents is aimed at engaging the parents to play a significant role in promoting STEM education to their children. Enculturing STEM learning in their children would involve the parents in assisting their children in STEM learning and motivating their children to venture into STEM careers. The promotion of parents' awareness of STEM education is through STEM-based reality television programmes. The game show requires players to have a sound understanding of STEM knowledge to win the game. This programme is aired on a national/private TV channel and viewed by the public and aims to increase parents' awareness of STEM education.

DEVELOP NATIONAL STEM STRATEGIC ACTION PLAN 2018–2025

To strengthen the country's STEM ecosystem, the MOE, MOSTI, and MOHE have developed the National STEM Strategic Action Plan 2018–2025. The action plan outlines seven focus areas, driven by 13 strategies and 33 initiatives. The seven focus areas are:

- Fostering awareness
- Improving the quality of teaching and learning
- Strengthening infrastructure and facilities
- Cultivation of research, development, innovation, commercialisation, and entrepreneurship
- Expansion of career opportunities
- Data harmonisation
- Strengthening strategic cooperation

This action plan complements the existing national policy related to education and science and technology policies, such as MEB 2018–2025, National Science, Technology and Innovation Policy (DSTIN) 2021–2030. The action plan aims to:

- Increase student interest, awareness, and enrollment in STEM
- Prepare and equip talent to meet future industrial and employment demands
- Produce competent human capital and help to enhance the country's competitiveness
- Make STEM part of the national science movement

As part of the National STEM Strategic Action Plan 2018–2025 and MEB 2013–2025, STEM education initiatives aim to increase the number of students who can think logically, use technology, solve problems, and be innovative. The MOE works with other government agencies, the private sector, and non-government organisations, to roll out a national public awareness campaign to educate the public (students, parents, educators, school counsellors, community, industry, etc.) on the importance of STEM and create an awareness of career opportunities in STEM-related fields

Reflection: From policy to implementation

The policies and strategies discussed were targeted at the implementation level and would most probably lead to the successful enactment of the policies, namely at the classroom level. The focus of various strategies was developing the teachers' competencies, awareness, knowledge, and pedagogical skills in STEM education. Nevertheless, empirical studies through bibliographic research (Cao et al., 2020) highlighted that researchers from higher institutions in Malaysia focused more on student outcomes. A review of Integrated STEM education (Bryan & Guzey, 2020) at the global level also points to the lack of studies on how teachers could execute a meaningful STEM teaching and learning environment. Thus, the effectiveness of the programmes and interventions advocated by the STEM policies stand to be proven. While students' outcomes could be a major indicator, it would be difficult to assess the effectiveness of the intervention without a controlled environment of the interventions. Most of the integrated STEM education in nature is often through non-formal learning and outreach programmes. Thus, critical evaluation of any educational interventions, including STEM competitions, campaigns, in-service training, and STEM learning materials is paramount. The outcome of student' learning serves as an evidence-based input in analysing the effectiveness of the policies.

Any polices successful implementation might be hindered by an overloaded policy (Bush et al., 2019). In the case of policies related to STEM education, both new pedagogical and subject matters are introduced. Such wide-ranging strategies and interventions might create the perception that there are many policy changes and eventually lead to ineffective implementation of the policies.

In the MEB document, strengthening the delivery of STEM education was the focus. Little information was provided on how assessment change will meet the changes in the delivery of integrated STEM education. Classroom-based assessment is beginning to occur in the Malaysian education system, intending to promote formative assessment and reduce the practice of summative assessment – a practice that is synonymous in an exam-oriented country. While classroom-based assessment can address students' soft skills and quality of artefacts from project-based STEM-related activities, less is known about the extent to which assessments of STEM education outcomes at the standardised examinations are made available. As it is commonly known, in a country that is examination-oriented, such as Malaysia, teachers teach to the test. Thus, teachers tend to implement few hands-on projects and problem-based activities at the classroom level.

Various systematic reviews on STEM education (Bryan & Guzey 2020; Kanadli, 2019) often indicate that conceptualisation of STEM education varies from teaching STEM as a single discipline to teaching STEM in an integrated manner. The ill-defined concept of STEM education in Malaysia also exists (Halim, 2018). The IBSE approach to STEM education emphasises that STEM education, espoused by the MOE, reinforces the nature of science and pedagogical inquiry practices. Inquiry approach is one of the pedagogical approaches that support the implementation of project-based teaching and learning activities. However, the IBSE approach also suggests that the teaching and learning activities implemented in STEM education portray the monodisciplinary approach rather than the integrated STEM approach. This dual approach signifies that Malaysia is still aiming to simultaneously fulfill the 60:40 policy and STEM education policies. The 60:40 policy emphasises enhancing the number of the students' participation in science at the upper secondary level and eventually participate in science disciplines (which are normally monodisciplinary) at the higher education. Thus, 60:40 focuses on the teaching of monodisciplinary subjects.

A way forward is to view STEM education in Malaysia as an integration of engineering practices with science and mathematics content and practices. In other words, the engineering practices bind together science and mathematics content and practices, including integrating the different disciplines in a meaningful manner to create technologies for a specific purpose (Bryan et al., 2016; Moore et al., 2014). The interdisciplinary approach offers students a chance to bring together ideas from STEM disciplines to solve real-world problems. As such, we believe that the construction of a deep understanding of STEM knowledge is necessary. To help students progressively develop critical insight, as suggested by Chalmers et al. (2017), teachers should prioritise a smaller number of conceptually larger, transferable ideas – big ideas of each individual STEM discipline as well as cross-discipline big ideas. Cross-discipline big ideas are concepts that have application across two or more domains of STEM disciplines, such as variables, patterns, models, computational thinking, reasoning, and argument (Chalmers et al., 2017).

Integrated STEM education would also see students' immersion in STEM professionals' practices (such as ask questions, define problems, plan and carry out investigations, and develop models). Engaging students in STEM professionals' practices help students understand how STEM professionals work and the nature of STEM discipline enterprises. For example, engagement in modelling and in critical and evidence-based argumentation help students understand that STEM professionals develop models to describe, test, and predict phenomena and design systems. This contributes to the development of a basic understanding that "science is a human endeavour" (an understanding about the nature of science by NRC, 2012) and "knowledge produced by science is used in engineering and technologies to create products to serve human ends" (a big idea *about* science by Harlen, 2015), and thus contribute to STEM literacy. Some characteristics of STEM-literate students as described by Bybee (2013) are "understanding of the characteristic features of STEM disciplines as forms of human knowledge, inquiry, and design" and "awareness of how STEM disciplines shape our material, intellectual, and cultural environments". In this regard, teachers should have an adequate or informed conception of these big ideas to help students understand individual and cross-discipline big ideas successfully. Further research and development is necessary to help teachers develop these big ideas and evaluate teaching for big ideas. Additionally, a research-based integrated STEM framework or curriculum is essential to help teachers design integrated unit plans and assessments.

Conclusion

The Malaysian MOE has systemically planned STEM Education policies and strategies to address STEM human resource building, to be at the top one-third of TIMSS and PISA ranking. The efforts are commendable, and initiatives such as SSCI+ are an example of implementing the policies with a contextual slant rather than a top-down implementation process. The reflection of policy to implementation offers a platform to identify strengths, weaknesses, and opportunities in implementing and monitoring the STEM education policies. Together with evidence-based effectiveness measures of the policies and implementation strategies, Malaysia will be able to improve the quality of STEM education delivery and eventually realise the goals of the policies.

References

Bahagian Pembangunan Kurikulum (BPK). (2014). *Kemahiran berfikir aras tinggi: aplikasi di sekolah*. Putrajaya: Kementerian Pendidikan Malaysia.

Bahagian Pendidikan Guru (BPG). (In press). *STEM Teachers' Competency Framework*. Putrajaya: Kementerian Pendidikan Malaysia.

Bryan, L., & Guzey, S. S. (2020). K-12 STEM education: An overview of perspectives and considerations. *Hellenic Journal of STEM Education*, *1*(1), 5–15.

Bryan, L. A., Moore, T. J., Johnson, C. C., & Roehrig, G. H. (2016). Integrated STEM education. In C. C. Johnson, E. E. Peters-Burton, & T. J. Moore (Eds.), *STEM Road Map: A Framework for Integrated STEM Education* (pp. 23–37). New York, NY: Routledge.

Bush, T., Ng, A. Y. M., Wee, K. T., Chay, J., Glover, D., & Lei, M. T. (2019). *Educational Policy in Malaysia: Implementation Challenges and Policy Proposals*. Singapore: The HEAD Foundation.

Bybee, R. W. (2013). *The Case for STEM Education: Challenges and Opportunities*. Arlington: NSTA Press.

Chalmers, C., Carter, M., Cooper, T., & Nason, R. (2017). Implementing "big ideas" to advance the teaching and learning of science, technology, engineering, and mathematics (STEM). *International Journal of Science and Mathematics Education*, *15*(1), 25–43.

Cao, T. H., Trinh, T. P. T., Nguyen T. T., Le, T. T. H., Ngo, V. D., & Tran, T. (2020). A bibliometric review of research on STEM education in ASEAN: Science mapping the literature in Scopus database, 2000 to 2019. *EURASIA Journal of Mathematics, Science and Technology Education*, *16*(10), 1–12. doi: https://doi.org/10.29333/ejmste/8500.

Davies, A., Fidler, D., & Gorbis, M. (2011). *Future Work Skills 2020*. Palo Alto, CA: Institute for the Future.

Deloitte Touche Tohmatsu Limited. (2016). *Talent for Survival: Essential Skills for Humans Working in the Machine Age*. London: Deloitte LLP.

Harlen, W. (2015). *Working with Big Ideas of Science Education*. Retrieved from: http://www.interacademies.org/publications/26703.aspx

Halim, L. (2018). STEM education: Issues and way forward. In *STEM Education in Malaysia* (pp. 37–59), Abd Manaf, Z., & Talib, Z. A. (Eds.), Malaysia: Management publication Malaysia Citation Centre Department of Higher Education Ministry of Higher Education.

Kanadli, S. (2019). A meta-summary of qualitative findings about stem education. *International Journal of Instruction*, *12*(1), 959–976. doi: https://doi.org/10.29333/iji.2019.12162a.

Ministry of Education (MOE). (2013). *Malaysia Education Blueprint 2013–2025*. Putrajaya: Kementerian Pendidikan Malaysia.

Ministry of Education (MOE). (2018a). *Quick Fact 2018: Malaysia Educational Statistics*. Putrajaya: Educational Planning and Research Division, Ministry of Education Malaysia.

Ministry of Education (MOE). (2018b). *2018 Annual Report: Malaysia Education Blueprint 2013–2025*. Putrajaya: Educational Planning and Research Division, Ministry of Education Malaysia.

Ministry of Science, Technology and Innovation (MOSTI). (2018). *Pelan Tindakan Strategik STEM Nasional 2018–2025*. Putrajaya: Kementerian Sains, Teknologi dan Inovasi.

Mohd Shahali, E. H., Halim, L., Rasul, M. S., Osman, K., & Mohamad Arsad, N. (2019). Students' interest towards STEM: A longitudinal study. *Research in Science & Technological Education*, *37*(1), 71–89. doi: https://doi.org/10.1080/02635143. 2018.1489789.

Mohd Shahali, E. H., Halim, L., Rasul, M. S., Osman, K., & Zulkifeli, M. A. (2017). STEM learning through engineering design: Impact on middle secondary students' interest towards STEM. *EURASIA Journal of Mathematics Science and Technology Education*, *13*(5), 1189–1211. doi: https://doi.org/10.12973/eurasia.2017.00667a.

Moore, T. J., Glancy, A. W., Tank, K. M., Kersten, J. A., Smith, K. A., & Stohlmann, M. S. (2014). A framework for quality K-12 engineering education: Research and development. *Journal of Pre-College Engineering Education Research*, *4*(1), 1–13. doi: https://doi.org/10.7771/2157-9288.1069.

National Research Council (NRC). (2012). *A Framework for K-12 Science Education: Practices, Crosscutting Concepts, and Core Ideas*. Washington, DC: The National Academies Press.

Shahali, E. H. M., Ismail, I., & Halim, L. (2017). STEM education in Malaysia: Policy, trajectories and initiatives. *Asian Policy Research*, *8*(2), 122–133.

3 A journey towards STEM education in Bhutan

An educational review

Kinley, Reeta Rai, and Sherab Chophel

Introduction

STEM education has led to educational reforms around the world as it has been argued to provide students with opportunities to imbue 21st-century skills and dispositions such as problem solving, critical thinking, creativity, curiosity, decision making, leadership, and entrepreneurship. STEM education prepares youth to become scientifically and technologically literate, so that they may become more employable and contribute productively to the nation's economy and social well-being (Zollman, 2012). It is believed that through STEM education, students develop skills such as problem solving, critical thinking, creativity, curiosity, decision making, leadership, and entrepreneurship. We assume educating youths in STEM fields can potentially bring tremendous growth to the nation's economy and social well-being. Since STEM education is linked to the development of human skills and national economic growth, it has become a major focus of education policies around the world (Montgomery & Fernández-Cárdenas, 2018). In recent years, increased emphasis on STEM education has encouraged educators and policymakers to promote STEM teaching and learning in classrooms, including the tiny Himalayan kingdom of Bhutan. In this chapter, we refer to STEM as an integration of two or more of the science, technology, engineering, and mathematics disciplines.

Bhutan is known as "Drukyul", the Land of Thunder Dragon. It is a landlocked mountainous country with a total area of 38,394 square kilometers (km²) located along South Asia's eastern Himalayas. It is bordered by India on the east, west, and south, and the People's Republic of China on the north. About 71% of the country is covered with forests, and a minimum of 60% of the country's total land shall be maintained under forest cover for all time (Royal Government of Bhutan [RGoB], 2008). Famed as one of the ten biodiversity hotspots in the world, the country is home to some of the rarest animal, plant, and bird species. The country has pledged to remain a net carbon sink in perpetuity (Powdyel, 2014).

Bhutan's current national educational structure consists of seven years of primary education (Pre-Primary (PP)–Class VI), two years of lower (Class VII–VIII), middle (Class IX–X), and higher secondary (Class XI–XII) schooling

DOI: 10.4324/9781003099888-3

Table 3.1 Structure of education system in Bhutan

Key stages	Grades	Age of students	Education programmes
Key stage V	Class XI to XII	16 to 17 years	Higher secondary education
Key stage IV	Class IX to X	14 to 15 years	Middle secondary education
Key stage III	Class VII to VIII	12 to 13 years	Lower secondary education
Key stage II	Class IV to VI	9 to 11 years	Primary education
Key stage I	Pre-Primary (PP) to Class III	5 to 8 years	Primary education
	ECCD early	3 to 5 years	Early childhood care and development pre-school program

(see Table 3.1), followed by university education within the country or abroad. Progression from one level to the next necessitates certification of their performance.

Over the years Bhutanese students' performance in mathematics and science has generally not been very encouraging across all the classes and, in particular, Classes X and XII. The subject-based item analysis for Class XII examination papers for 2018, 2019, and 2020 carried out by Bhutan Council for School Examination and Assessment (BCSEA) showed that students' poor performance in science subjects was attributed to their failure to understand certain concepts in different science subjects (BCSEA, 2018; 2019a; 2020). Their mathematics achievements were affected by their science performance due to the inter-dependency of the two subjects.

The Programme for International Student Assessment for Development (PISA-D) exam, which was held in 2017 for students (2,457) aged 15 years in Class VII and above to benchmark Bhutanese students' competence in science and mathematics literacy at the international level, found Bhutanese students to be below average, with scores of 45.10% and 38.84%, respectively (BCSEA, 2019b). The PISA-D report calls for strengthening and improving competency-based activities and assessment in the curriculum, and it instructs teacher-training colleges to include competency based teaching and learning modules in their pre-service training programs.

A study by Childs et al. (2012) contended that some of the issues and challenges faced by science education in Bhutan included inadequate content and pedagogical knowledge of teachers, and the fragmentation of the current science curriculum. Further, Bhutanese students' declining interest in science and mathematics subjects may also be attributed to the compartmentalised approach to teaching science subjects in Bhutanese schools. Mathematics is a compulsory subject for children despite their low levels of competency and the lack of interest in mathematics (Rinzin, 2018). The use of technology to assist teaching and learning processes also remained minimal. Barriers such as a lack of appropriate resources had hindered the successful integration of technology for teaching and learning in Bhutanese schools.

Hence, there is a need to change the ways mathematics and science are currently taught in the schools and higher education institutions of Bhutan. Students' fluency in mathematics and science is much needed for building a scientifically literate society, enhancing the country's economy, mitigating employment problems, and other challenges faced by Bhutan. In the next section, we provide an account of the emergence and evolution of modern education in Bhutan and Bhutan's education system, before providing an overview of the science and STEM education in Bhutan.

Background on the modern education system in Bhutan

The journey of education in Bhutan started with monastic education before the institution of the modern education system. Prior to the 1960s, literacy was confined to the monasteries where the social, economic, and spiritual needs of the country were served by monastic education. The sources of education in the monastic system included the study of the Buddhist religion, literature, astrology, philosophy, fine arts, theology, medicine, architecture, and music (Gyamtsho & Drukpa, 1998). The first modern education in Bhutan began in 1914 with the establishment of the first modern school in Haa (Dorji, 2016). The beginning of the journey of modern education was marked by sending 46 boys to India for Western education in 1914 (Tobgye, 2018). Modern education was established with the help of India which allowed Bhutan to borrow the entire education system with English as the medium of instruction along with teachers, curriculum, and teaching-learning materials (Zangmo, 2018). As such the education system in Bhutan had colonial relics from India. However, during that time, the system was deemed relatively young and formal schooling in the government system only started in the 1950s. The main subjects taught were Hindi (the formal language of India), English, and Arithmetic, along with Dzongkha. Until the 1960s, the subjects were mostly taught in the medium of Hindi, partly in English, and of course in Dzongkha. In 1962, English was made the main medium of classroom instruction in the schools of Bhutan to enable the small and isolated country to communicate with the world around it in the future.

The education system in Bhutan evolved over the years with the introduction of Five-Year Development Plans. With the launch of the first Five-Year Plan in 1961, modern education was further strengthened under the visionary leadership of Third Druk Gyalpo, King Jigme Dorji Wangchuck, known as "the father of modern Bhutan", to develop national human resource capacity and drive the socio-economic development of the country. The king established English medium schools across the country in addition to the approximate 11 schools with 400 students that existed in 1961 (Ministry of Education [MoE], 2014). The number of schools, students, and teachers had exponentially scaled up over the years since 1961.

Modern/Western education in Bhutan began in 1961. Compulsory formal education is provided through general school education to create an avenue for the citizens of Bhutan to embrace and provide more dimensions of education apart from monastic education. The formal school system consists of a year of PP education, followed by six years of primary education, two years of lower secondary, two years of middle secondary education, and two years of higher secondary education. Then it is followed by various tertiary levels of education. Despite the late start, school education in Bhutan has witnessed unprecedented growth and progress within a period of over six decades. The modern education system has expanded from about 11 schools prior to the 1960s to 1,132 schools and other educational institutes in 2020, spanning from early childhood care education to tertiary and technical and vocational education (MoE, 2020a).

Science education in Bhutan

In Bhutan, science education started with the inception of Western education in 1961. Over the years, the science curriculum had to undergo several changes and reformation at different stages in different ways because it had been introduced without a proper plan in the early days (Childs et al., 2012). Until the 1980s, Bhutan had been using the borrowed science curriculum from neighbouring country, India. In view of making the curriculum content for PP to Class IV aligned with the learning needs of children in Bhutan, the Education Department introduced a system called New Approach to Primary Education (NAPE) in 1985. With the new approach, a new subject called Environmental Studies (EVS) was introduced in place of social studies and science (Dorji, 2016). The NAPE was an innovative curriculum that required teachers to adapt to a new set of curriculum materials and a new approach to teaching that was oriented towards child-centred learning. The NAPE sought to make the primary science curriculum in Class IV–VI take more account of teaching and learning through Bhutan's natural and social environment and placed more emphasis on the development of investigative skills in children.

In 1999 and 2000, three distinctive science disciplines (physics, chemistry, and biology) for Classes VII and VIII were replaced by a single integrated science curriculum that emphasised learning science through the environment (Zangmo, 2016). Thus, the curriculum from Class IV to VIII became an integrated science curriculum, in which the learning was based on students' immediate environment and real-life experiences. For PP to Class III, science was studied through integrated EVS, in which students were expected to develop awareness, curiosity, and appreciate their immediate living and non-living environment around them. However, the science curriculum for Classes IX to XII remains similar to the single science curriculum borrowed from India.

The government's 10th Five-Year Plan (for the period 2008–2012) prioritised the reform of the science curriculum. A needs assessment study was carried out to identify and understand the issues and challenges presented by the number of revisions the science curriculum had undergone and collected well-informed perspectives from different stakeholders (teachers, principals, students, college lecturers, and science professionals). The findings of the needs assessment study, in general, revealed that the science curriculum in Bhutan was fragmented, lacking coherence and progression of ideas across grades. The findings also revealed that science practicals were taught separately without integrating with theory. There was also a general public perception on the decline of science standards that the science curriculum did not prepare learners for the world of work and scientifically sound citizens.

Based upon the needs assessment study, a major science curriculum reform as part of the 10th Five-Year Plan took place from PP to Class XII. The reformed curriculum is a spiral curriculum that builds a better foundation in learning science. Moreover, a STEM curriculum is also in place, which is an idea of educating students in four specific disciplines – science, technology, engineering, and mathematics – in an interdisciplinary and applied approach (MoE, 2020a). The study of STEM is identified as one of the essential learning areas in the Bhutanese curriculum that would provide innovative and creative skills under the science and technology education required for the 21st century. The STEM curriculum would prepare graduates with sound knowledge and skills in science and technology which would fulfill their personal needs, address societal issues, create career opportunities, and excel academic and professional opportunities in science and mathematics.

Information and technology in Bhutan

The first telephone network in Bhutan was established in 1963. Since then, the telecommunication network has evolved from a physical wire network to the national digital network that is in existence today. Today, telecommunication is given high priority in Bhutan's development plans. The first international satellite link was established in 1990 with the installation of the Earth Satellite Station and an international gateway switch in Thimphu, the capital city of Bhutan. With the fear of outside influence for years, the country remained in isolation, and it was only in 1999 that the country established its first television network and also started Internet services (BBC, 2019). By March 1999, all of the 20 district headquarters had access to telecommunication services. Since then, information and communication technology (ICT) has provided access to global information and plays an important role in the country. Nevertheless, large sections of the population have yet to adopt ICT and to appropriate it for their own purposes (National Statistics Bureau [NSB], 2013).

Recognising the implication and power of technology in transforming and enhancing the quality of education, the MoE developed a very comprehensive document "the Bhutan isherig-2 education ICT masterplan" that envisions supporting "Nationally rooted and globally competent citizens through the equitable and pervasive use of emerging and relevant technology which is aimed to leverage the power of ICT to enhance the quality of education" (MoE, 2019, p. 1).

The outbreak of the COVID-19 pandemic has affected livelihood and education worldwide. With no exception, Bhutan also experienced the impact and made paradigm shifts in teaching, learning, and assessments. Schools and higher education institutions had to deliver lessons online. The significance of ICT became more evident. There was an urgency to harness the affordances of ICT in education in schools and higher education institutions of Bhutan to ensure that teaching and learning continued to take place. In his interactions with teachers who were undergoing ICT training, the Prime Minister of Bhutan, Dr. Lotay Tshering, said, "Besides Dzongkha and English, ICT has to be the third primary subject that students need to compulsorily pass to move onto the next grade" (Tshedup, 2020). Likewise, His Majesty The King of Bhutan, King Jigme Khesar Namgyel Wangchuck symbolically handed over to the people of Bhutan two *kashos* (royal edict) on the reforms decreed for the education and civil service on December 17, 2020. The following is an excerpt from his address on education reform highlighting ICT:

> In preparing our youth for the future, we must take advantage of available technologies, adapt global best practices, and engineer a teaching-learning environment suited to our needs. Technology is the argument of our time and a major indicator of social progress. The irony in our context is the absence of technology in classrooms for a generation of students who are exposed to, and live in the digital age. To ensure that teachers are not disconnected from their students, professional development of teachers should integrate technology, digitalisation, artificial intelligence, and automation.
>
> (Kuensel [translated], 2021)

The emphasis on ICT has implications for STEM education in Bhutan. In the next section, we provide an account of STEM education plans and policies in Bhutan.

Introduction to Bhutan's current STEM education system

STEM education plays a pivotal role in nation-building. As a developing country, Bhutan places great importance on institutionalising a relevant and challenging STEM curriculum for all of its school-going children. The Bhutan Education Blueprint 2014–2024 (MoE, 2014) states succinctly that school curricula must foster the acquisition of the 21st-century skills of innovation, creativity, enterprise, and universal human values of peace and harmony. STEM education, which has emerged in educational reforms around the world, is widely perceived as a means for students to acquire such knowledge, skills, and values.

The current Bhutanese education system (BES) provides segregated domain-specific STEM education. From the acronym STEM, only science (S) and mathematics (M), and ICT are mandatory core curriculum of the BES from primary to middle secondary level (Class X). Vocational technology is taught in some selected schools as an optional subject from Classes IX to XII. Similarly, the BES does not offer engineering as a subject in the school curriculum, although some aspects of engineering knowledge, such as problem solving and innovation, are encouraged in science and mathematics subjects. Furthermore, Agriculture for Food Security (AgFS) is offered as an optional subject in some schools from Classes IX to XII. The AgFS curriculum aims to reduce unemployment and social issues, as well as maintain the national food security index and create pathways to higher education. Health and physical education (HPE) is taught to students in Classes PP to VI to provide health and physical literacy education for leading healthy lifestyles. EVS is introduced to students in Classes PP to III, and environmental science is offered as an optional subject in Classes IX and X.

From Classes PP to XII, there are separate curriculum frameworks with specific strands for all of the aforementioned STEM subjects. The Royal Education Council (REC), the Ministry of Education (MoE), and the Bhutan Council for School Examinations and Assessment (BCSEA) ensure the highest standards in the selection and organisation of STEM content into curriculum frameworks, as well as the use of appropriate pedagogies (e.g., Kagan's cooperative learning strategies, place-based education, flipped classroom, etc.) and assessment strategies.

STEM education at primary level key stage I (4 years)

A child who turns five years old on or before February 5th is admitted to class PP in both public and private schools. From Classes PP to III, the mandated curriculum includes English, Dzongkha, and mathematics, while HPE, arts education, value education, and ICT literacy are compulsory but not assessed. In the past, science was taught as a general science subject from Classes PP to VI. With the introduction of the NAPE in the 1980s, science in Classes PP to III was taught as part of EVS in Dzongkha. Through EVS, students in key stage I are introduced to fundamental science concepts such as living and non-living things, plant and animal names, and so on. At this stage, students are expected to develop awareness, curiosity, and knowledge about their immediate living and non-living environments.

Key stage I mathematics is taught using materials found in the students' immediate surroundings, based on the constructivist learning approach. The mathematics curriculum emphasises mathematical learning in five major areas, namely, numbers and operations, patterns and algebra, measurement, geometry, and data and probability. Since mathematics is taught in English, learning mathematics in English is difficult for Bhutanese primary students, particularly those in Class PP (Dolma, 2016). At this stage, the HPE curriculum focuses on the development of fundamental movement skills, water sanitation and hygiene (WASH) practices, social skills, healthy dietary habits, and active living (REC, 2021). Formative assessments are used for students in Classes PP to II, while students in Class III take a year-end competency based assessment test (CBAT) in English, Dzongkha, and mathematics administered by the BCSEA.

STEM education at primary level key stage II (3 years)

From Classes IV to VI, the mandated curriculum includes English, Dzongkha, science, mathematics, and social studies, while HPE, arts education, value education, and ICT literacy are compulsory but not assessed. Students in Classes IV through V are evaluated using 50% formative and 50% summative assessments administered by the respective school authorities while students in Class VI take a year-end CBAT administered by the BCSEA.

During key stage II, science is introduced to Class IV students for the first time. Students learn about a wider variety of living things, materials, and scientific phenomena occurring in their surroundings. They gradually begin to connect ideas, explain things using simple models and theories, and conduct scientific investigations.

STEM education at lower secondary level key stage III (2 years)

The core curriculum in key stage III includes English, Dzongkha, mathematics, science, history, and geography. The mathematics strands include numbers and operations, patterns and algebra, geometry, and data management and probability. Until 1998, science was taught in Classes VII and VIII as three separate subjects, namely, physics, chemistry, and biology. These three science subjects were replaced by locally contextualised integrated science. There was one science textbook containing chapters from physics, chemistry, and biology. Integrated science textbooks titled "Science for Class VII: Learning Science through Environment" and "Science for Class VIII: Learning Science through Environment" were implemented across the country in 1999 and 2000, respectively.

The integrated science for Classes VII and VIII was introduced in an attempt to create a truly Bhutanese science for students but the teachers in the field found it difficult to teach this subject because they lacked expertise in all three sciences. According to Sherpa (2007), introducing an integrated curriculum without proper research was either inappropriate or a hasty decision that confused teachers and impacted students' science learning. Logistically, integrating the three disciplines of science in these two classes was intended to alleviate the science teacher shortage in schools, as only one teacher would be required to teach science in one grade or class section (Tenzin & Lepcha, 2012). Although students were excited about the new curriculum, because it provided minimal cognitive challenges and allowed students to construct scientific ideas from their surroundings, the implementation of integrated science in the field was confronted with several challenges. According to Childs et al. (2012), Class VII integrated science included more chemistry concepts, whereas Class VIII was more biology focused. There was also a disparity in the inclusion of topics and contents, resulting in a weak foundation for the transition to single sciences in Class IX and beyond.

In 2007–2008, the MoE's Department of Curriculum Research and Development (DCRD) conducted an in-depth needs assessment of science education in Bhutan with assistance from UNESCO in New Delhi. This study revealed a pressing need to revise the integrated science curriculum for Classes VII and VIII. As a result, a study on the relevance of integrated science was conducted in 2011 and 2012 by Tenzin and Lepcha (2012). According to this study, the integrated science content was more appropriate for primary level education but it did not lay a strong scientific foundation for higher levels of learning. Similarly, teachers were not professionally or academically prepared to teach integrated science. Due to the aforementioned drawbacks, integrated science was replaced in 2017 by general science textbooks with three strands of science learning, namely, life processes, materials and their properties, and physical processes.

STEM education at middle secondary level key stage IV (2 years)

At the middle secondary level, science and mathematics are mandated core subjects, while students have the option of choosing either ICT or environmental science as an optional subject. Since 2011, five schools located closer to technical training institutes have been also offered TVET (Technical and Vocational Education and Training) courses as an optional subject for Classes IX and X. This initiative was taken to promote vocational programs starting from school education. Under the vocational skills curriculum, students are taught general electrical and house wiring, basic automobile engineering, carpentry, plumbing, tailoring, hardware training, and Zorig Chusum (traditional arts and crafts of Bhutan).

The science curriculum is divided into three disciplines – biology, chemistry, and physics. During key stage IV, students are expected to develop fundamental knowledge and understanding of the concepts of chemistry, biology, and physics, as well as the fundamental skills required for their application in new and changing situations. Many students choose to pursue higher education in science and mathematics based on their understanding, process skills, and attitude toward the subjects. Students wishing to study science with or without mathematics in Class XI must pass the national board exam and meet the set cut-off points to be admitted to the science stream.

Students are introduced to higher and deeper concepts of science and mathematics beginning in Class IX, but there have been reports of students underperforming at key stage IV. The successive pupil performance reports for Class X (Table 3.2) show that mean science and mathematics performance is below average (BCSEA, 2018, 2019a, 2020).

Students' poor performance in mathematics affects their performance in science as well because excelling in science requires a strong foundation in mathematics. Mathematics is a required subject for admission to Class XI in the Pure Science and Commerce streams. Students interested in the science stream must

Table 3.2 Subject-specific percentage passes (%) and mean score (M)

Subject		2019 (N = 12614)	2018 (N = 12462)	2017 (N = 11973)
Science	Physics	Pass: 89.24%	Pass: 95.23%	M = 56.05
		M: 50.34	M: 55.47	
	Chemistry	Pass: 79.19%	Pass: 84.55%	M = 48.43
		M: 45.84	M: 46.89	
	Biology	Pass: 95.73%	Pass: 94.37%	M = 45.73
		M: 53.29	M: 53.20	
Mathematics		Pass: 88.44%	Pass: 97.01%	Pass: 91.84%
		M: 52.02	M: 56.83	M: 49.77

Source: BCSEA.

achieve a minimum of 40% in mathematics and a minimum of 55% in science, with passing marks in biology, chemistry, and physics. Similarly, the merit order for students choosing commerce would be based on their mathematics marks, with each student needing a minimum of 40%. Many students are unable to qualify for Class XI due to poor mathematics performance, so they drop out of Class X and enter the labour force. Since mathematics is a mandatory curriculum at key stage IV, students are compelled to learn despite their unwillingness and lack of interest in mathematics (Rinzin, 2018).

STEM education at higher secondary level key stage V (2 years)

After completing middle secondary education, students continue their education for two years at higher secondary education. Students at key stage V study either bioscience (physics, chemistry, and biology) or pure science (physics, chemistry, and mathematics). Some students study both mathematics and biology in their bid to receive equal opportunities in both fields for undergraduate programs. Along with the mentioned mathematics and science curricula, English and Dzongkha are mandatory. Students who excel academically at the national board exam, Bhutan Higher Secondary Education Certificate (BHSEC), are sent to foreign countries to further their education in STEM fields such as medicine, biomedical, engineering, and information technology. Every year, around 150 academically outstanding students receive scholarships from the Royal Government of Bhutan (RGoB). Further, to improve the quality of secondary STEM teachers, 5% of scholarships are awarded to outstanding students to pursue bachelor degrees in physics, chemistry, biology, and mathematics in first-world countries such as the United States of America, Canada, and Australia, on the condition that they join the teaching cadre after completing their course. The Scholarship and Student Support Division of the MoE constantly monitors the academic progress of the earmarked scholars. All scholars must return to Bhutan and take the competitive Royal Civil Service Examination (RCSE) and only the successful students get assigned to various ministries. Students who do not qualify for international scholarships study in colleges affiliated with the Royal University of Bhutan (RUB) in a variety of STEM fields. So far, students who have received international scholarships have been a source of manpower, particularly in the fields of medicine, ICT, and engineering.

STEM education during the COVID-19 pandemic

On March 18, 2020, all schools in Bhutan closed due to the COVID-19 pandemic. This resulted in 170,263 children from Classes PP to XII continuing their education from home. This precautionary move was undertaken in response to fears that students would become infected, and hence, closing schools would prevent rapid local transmissions. In addition, schools lacked adequate WASH infrastructure and facilities. As such, the MoE implemented a national plan for Education in Emergency (EIE) to ensure the continuity of education.

Around 2,770 volunteer teachers from schools formed an organisation called Volunteer Teachers of Bhutan (VTOB) to work on student engagement during the pandemic (Tashi, 2020). An eLearning program was launched, and volunteer teachers began broadcasting lessons on the national television channel, Bhutan Broadcasting Service (BBS), as a form of virtual education. Students would tune in daily on BBS and attend lessons at the scheduled times designed based on the key stages. The curricula were also modified and prioritised to achieve the desired learning outcomes (MoE, 2020b). In addition to the eLearning Programme, students from key stages II to V were engaged by their subject teachers through a variety of online tools such as Google Classrooms, Wechat, e-textbooks from the REC, free access to *eKuensel* (national newspaper of Bhutan), free e-books, and many more. Students in remote areas were given self-instructional materials (SIM) that were designed in accordance with EIE. Students across the country were promoted to the next grade level based on competency based formative assessments, except for those taking national board examinations in Classes X and XII.

Although the pandemic continues throughout the world, including Bhutan, all schools opened in February of 2021, and students are taught using the New Normal Curriculum (NNC). In contrast to conventional education, where teachers teach students what to learn, what to do, and what to value, NNC would teach students how to learn, do, and value, based on competency-based learning. Unlike last year, the NNC is expected to ensure that there is no disruption in learning this year. The NNC instructional guides (NNC IG) direct the teacher's role as facilitators or guides in the transformation of textbook knowledge-based learning to process-based, competency based, and experiential learning (REC, 2021). The NNC IG employs blended learning and flipped classroom methods of learning to facilitate learning at any time and from any location while taking into account each learner's individual learning differences and situations.

Conclusion

In this chapter, we have provided a historical narrative of Bhutanese education with a special focus on science and mathematics education. The narrative suggests the evolving nature of STEM education in Bhutan. While smaller attempts of integrated science teaching have already been initiated in the past, the momentum and application are picking up as teachers and teacher educators become more aware and conversant with an integrated approach to STEM education. In current times, the motivation amongst educators has heightened upon drawing inspiration from the solemn decree from the King of Bhutan.

We would like to close this chapter in the words of His Majesty The King of Bhutan, King Jigme Khesar Namgyel Wangchuck, who symbolically handed over to the people of Bhutan two *kashos* (royal edict) on the reforms decreed for the education and civil service on December 17, 2020. The following is an excerpt from the address highlighting the need for educational reform and the importance of STEM subjects, translated by Kuensel:

> Therefore, our generation has the sacred responsibility of radically rethinking our education system and transforming curriculum, infrastructure, classroom spaces, and examination structures. Educationists and experts have identified what twenty-first century competencies mean for children everywhere. By developing their abilities for critical thinking, creative thinking, and learning to be life-long learners, we have to prepare them to be inquisitive, to be problem-solvers, to be interactive and collaborative, using information and media literacy as well as technological skills. We must prioritise self-discovery and exploration, and involve learners in the creation of knowledge rather than making them mere consumers of it. We must make STEM subjects part of their everyday language.
>
> (Kuensel, 2021)

The educational approaches adopted are rapidly expanding in Bhutan. This trend is now likely to be bolstered following the 2020 *kashos* for education reform. As stated by His Majesty, Bhutanese educators must revisit curriculum, pedagogy, and learning strategies to shift student practice away from passive modes of learning. New and revitalised educational practices must prepare students to thrive in the modern era, and embody such qualities as an inquisitive disposition and ability to problem solve. Therefore, the progressive move towards enhancing STEM education has become a key priority for Bhutan.

References

BBC. (2019, August 27). *Bhutan Profile – Media*. BBC. http://www.bbc.com/news/world-south-asia-12484025

Bhutan Council for School Examination and Assessment. (2018). *Pupil Performance Report 2018 (Vol. 11)*. Thimphu, Bhutan: Bhutan Council for School Examinations and Assessment.

Bhutan Council for School Examination and Assessment. (2019a). *Pupil Performance Report 2019 (Vol. 12)*. Thimphu, Bhutan: Bhutan Council for School Examinations and Assessment.

Bhutan Council for School Examination and Assessment. (2019b). *Education in Bhutan Findings from Bhutan's experience in PISA for Development*. Thimphu, Bhutan: Bhutan Council for School Examinations and Assessment.

Bhutan Council for School Examination and Assessment. (2020). *Pupil Performance Report 2020 (Vol. 13)*. Thimphu, Bhutan: Bhutan Council for School Examinations and Assessment.

Childs, A., Tenzin, W., Johnson, D., & Ramachandran, K. (2012). Science education in Bhutan: Issues and challenges. *International Journal of Science Education, 34*(3), 375–400. doi:10.1080/09500693.2011.626461.

Dolma, P. (2016). *Investigating Bhutanese mathematics teachers' beliefs and practices in the context of curriculum reform* [Doctoral dissertation]. Queensland University of Technology.

Dorji, J. (2016). International influence and support for educational development in Bhutan. In *Education in Bhutan* (pp. 109–124). Singapore: Springer.

Gyamtsho, D. C., & Dukpa, N. (1998). Curriculum Development for Primary and Secondary Education. *26*(1). Royal University of Bhutan. 70–76.

Kuensel. (2021, February 2). *Royal Kashos on Civil Service and Education*. Kuensel. https://kuenselonline.com/royal-kashos-on-civil-service-and-education/

Ministry of Education. (2014). *Bhutan Education Blueprint 2014–2024 Rethinking Education*. 1–173. Ministry of Education.

Ministry of Education. (2019). *iSherig 2-Education ICT Master Plan 2019–2023*. Ministry of Education.

Ministry of Education. (2020a). *Annual education report-2019–2020*. Policy and Planning Division, Royal Government of Bhutan.

Ministry of Education. (2020b, March 26). *Guidelines for Curriculum Implementation Plan for Education in Emergency (EIE)*. Thimphu: Royal Government of Bhutan. http://www.education.gov.bt/wp-content/uploads/2020/03/Guidelines-for-Curriculum-Implementation-Plan-for-Education-in-EmergencyEiE.pdf

Montgomery, C., & Fernández-Cárdenas, J. M. (2018). Teaching STEM education through dialogue and transformative learning: Global significance and local interactions in Mexico and the UK. *Journal of Education for Teaching, 44*(1), 2–13.

National Statistics Bureau. (2013). *Statistical Yearbook of Bhutan*. Thimphu: National Statistics Bureau, Royal Government of Bhutan.

Powdyel, T. S. (2014). *As I Am, so Is My Nation: A Tribute*. Thimphu: Kuensel Corporation Limited.

Royal Education Council. (2021). *New Normal Health and Physical Education Curriculum Framework Classes PP–XII*. Paro, Bhutan.: Royal Education Council, Royal Government of Bhutan.

Royal Government of Bhutan. (2008). *The Constitution of the Kingdom of Bhutan*. Thimphu: Bhutan.

Rinzin, K. (2018, December 15). *Education in Bhutan – Achievements and Challenges – Part 1 of 2*. Thimphu: The Bhutanese. https://thebhutanese.bt/education-in-bhutan-achievements-and-challenges-part-1-of-2/

Sherpa, A. (2007). Perspectives of secondary school science teachers on integrated science in Bhutanese schools. *RABSEL the CERD Educational Journal*, X, 5–33.

Tashi, S. (2020, March 25). *Volunteer Teachers Form Group to Help during COVID-19.* Thimphu: Business Bhutan. https://businessbhutan.bt/2020/03/25/volunteer-teachers-form-group-to-help-during-covid-19/

Tenzin, W., & Lepcha, S. (2012). *Relevancy of Integrated Science for Classes VII and VII in Bhutan: Understanding the Integrated Science in Bhutan – Concepts, Implementation and Challenges.* Paro: Department of Curriculum Research and Development (DCRD), Ministry of Education, Royal Government of Bhutan.

Tobgye, S. (2018, May 12). *The Unsung Heroes.* Thimphu: Kuensel. https://kuenselonline.com/the-unsung-heroes/

Tshedup, Y. (2020, February 4). *ICT Must Be the Third Language for Bhutanese: PM.* Thimphu: Kuensel. https://kuenselonline.com/ict-must-be-the-third-language-for-bhutanese-pm/

Zangmo, S. (2016). Grade 10 and 12 Bhutanese students 'attitudes toward science in the Thimphu district of Bhutan. *Journal of Turkish Science Education, 13*(3), 199–213. doi: https://doi.org/10.12973/tused.10180a.

Zangmo, Z. (2018). *Educational borrowing in the Bhutanese education system.* [Master's thesis]. Queensland University of Technology.

Zollman, A. (2012). Learning for STEM literacy: STEM literacy for learning. *School Science and Mathematics, 112*(1), 12–19. doi: https://doi.org/10.1111/j.1949-8594.2012.00101.x.

Section 2

STEM teacher education

Confluences of policy, professional learning, and practice

Ban Heng Choy

Introduction

Teaching any of the sciences, technology, engineering, and mathematics (STEM) disciplines as a subject on its own is challenging. What teachers know, believe, and care about teaching and learning is critical for improving the quality of teaching (Hattie, 2003). The challenge and complexity of teaching *integrated* STEM, in which the connections between two or more STEM disciplines are emphasised, is even more overwhelming for many teachers. Despite the important role of STEM teacher education to improve the quality of integrated STEM instruction, research on STEM teacher development is limited (Li & Anderson, 2020), and little is known about how best to prepare pre-service teachers for STEM-infused classrooms (Rinke et al., 2016). The ideas presented by Nivera, Limjap, Paderna, and Pastor and her colleagues from the Philippines (Chapter 4) as well as those by Faikhamta and Lertdechapat from Thailand (Chapter 5) are important contributions to address these gaps, particularly in the Asian context. In this commentary, I discuss how the Philippines and Thailand have responded to calls to reform their teaching and learning processes in response to the need to develop STEM competencies for the purpose of nurturing future-ready learners. More specifically, I will focus on how policies related to teacher education in these countries are related to their approaches in teachers' professional learning and enhancing teaching practices in the classroom. Here, I use the metaphor of *confluences* – where two or more rivers, each with their own flow and paths, meet to form a bigger river – to characterise how policies, professional learning approaches, and teaching practices come together or harmonise with the aim of providing high-quality STEM learning experiences. I argue that it is the confluences of policies, professional learning, and practices that position these countries in the right direction towards improving education in the STEM fields and illustrate these confluences using Singapore as an example.

DOI: 10.4324/9781003099888-II

Policy, professional learning, and practice

Improving STEM education is a multi-faceted enterprise that requires *coordination* across different policies and practices at various levels of engagement between multiple stakeholders. As argued by Fullan (2006), effecting changes in educational contexts has to go beyond simply implementing standards-based reform initiatives, professional learning communities or qualification frameworks focused on getting the "best teachers" or retaining the "best leaders" in the system. What is needed for reforms to work is to recognise that "any strategy of change must *simultaneously* focus on changing individuals and the culture or system within which they work" (Fullan, 2006, p. 7). While implementing STEM reforms may necessarily involve creating model STEM instructional units or practices, changing policies, building capacities through professional learning, and evaluating the impacts of these reforms (Bybee, 2010), it may require a rethink about the timeline of implementing these phases of reform. Likewise, changes to policy, professional learning, and practices for improving STEM teacher education cannot take place in silos or in a sequential manner. What the experiences of the Philippines and Thailand have taught us, as described in Chapters 4 and 5, is that these changes have to take place together and efforts are needed to "harmonize" these changes.

In the case of the Philippines, as described by Nivera et al. (Chapter 4), the country has embarked on an ambitious educational reform programme to address the challenges in teacher knowledge, teacher recruitment, teacher preparation, and teacher professional development since the 1990s. It is clear from the start that these challenges are interrelated and addressing them requires efforts that are interdependent. To this end, the Teacher Education Council (TEC) was created to be an umbrella organisation that serves to "harmonize plans, programs, and projects for teacher education" by strengthening and improving teacher education. Amongst the initiatives implemented, it is noteworthy that Centres of Excellence were established to lead in teacher education and support other teacher education institutions with the prioritised funding given by the government. Although there is no Centre of Excellence for STEM Education, these centres of excellence provide a platform to experiment with new ideas for teacher education with the aim of improving teacher quality as a whole. In addition, this initiative was coupled with the introduction of the Philippines Professional Standards for Teachers (PPST), which outlines the competencies and skills expected of all K-12 teachers. The standards form the basis of all professional learning activities in the Philippines.

A case in point is how the Philippine Normal University, which is the national centre for teacher education, has implemented innovative mathematics and science programmes to strengthen their teacher preparation programmes. These programmes include additional six units of pedagogical content knowledge courses and a three-unit summer science internship course for their pre-service teachers. As argued in Chapter 4, Nivera and her colleagues highlight the need for other teacher education institutes to rethink and redesign their programmes, as well as the need for stakeholders to provide curriculum flexibility and the necessary infrastructure for these institutes to initiate, develop, sustain, and evaluate these innovations. Besides overcoming issues related to pre-service teacher education, the authors also highlighted the need to re-examine the kind of training received by in-service teachers. In particular, there is a need to consider other kinds of programme structures beyond the "train-the-trainer" or cascading model to monitor and improve teacher professional development. All these initiatives worked in tandem with the push for better employment opportunities and higher salaries for teachers, as well as multi-sectoral efforts to attract high-quality high school graduates into STEM-related teacher education programmes through scholarships.

Teacher professional development has to evolve in sync with the emerging learning needs of teachers as they are confronted with an ever-changing educational landscape. To coordinate a more responsive approach to developing STEM teachers, Faikhamta and Lertdechapat in Chapter 5 recommend that teacher preparation, teacher professional development, and teacher education related policies should emphasise the idea of Pedagogical Content Knowledge (PCK) for STEM or PCK-STEM. Building on Shulman's (1987) ideas, PCK-STEM comprises orientations to teaching STEM, knowledge of STEM curriculum, knowledge of learners' understanding in STEM, knowledge of instructional strategies for teaching STEM, and knowledge of assessment of STEM learning (Lertdechapat, 2020).

As described in Chapter 5, the notion of PCK-STEM can be used to guide the design of pre-service and in-service science teacher education. What is noteworthy in these efforts is how these professional learning sessions are closely tied to teaching practices in the classroom. For example, the lesson study protocol was used in pre-service teacher education settings during student teaching to support student teachers in learning from the processes of lesson planning, lesson enactment, lesson observation, and post-lesson discussion (Lewis et al., 2011). These tasks in lesson study mirror the kind of teaching practices that occur in everyday classrooms. Similarly, a PCK-STEM focused professional development programme provides opportunities for teachers to share their understanding of STEM, articulate problems and issues in teaching STEM, develop an inquiry-based approach to improving teaching through action research, and apply new ideas in their own classrooms. Moreover, in both of these professional learning experiences, there is a deliberate effort to create learning opportunities within the operational contexts of pre-service and in-service teachers. This contrasts with the cascading model, as described in Chapter 4.

Despite the differences in approaches to improve STEM teacher education, both countries use some overarching ideas to frame their efforts. In the Philippines, the Teacher Education Council helps to coordinate efforts by the different stakeholders to improve policy implementation, professional learning and practice using the Philippines Professional Standards for Teachers as guidelines. In Thailand, the PCK-STEM articulates the kind of orientations and knowledge necessary for enacting classroom practices that are productive towards developing STEM competencies in their students. This set of orientations and knowledge forms the basis of all professional learning practices. Although the approaches in the two countries emphasise different aspects of policy, professional learning, and practice, what is common is the deliberate efforts to ensure some coherence between policies, professional development, and what happens in the classroom. To a large extent, these ideas from both countries reflect the features of an effective professional development programme, namely content-focused, active learning, coherence, duration, and collective participation situated within the teachers' context (Desimone, 2009). While it is not always clear what the impact of a STEM-focused professional development programme may be (Anderson & Tully, 2020), or whether a specialised STEM teacher preparation programme is useful (Teo & Ke, 2014); what is clear is that the two countries are moving in the right direction towards their own visions and goals of STEM education.

Confluences of policy, professional learning, and practice

The two chapters on STEM teacher education are good reminders for us, as STEM educators, that improving STEM teacher education is a journey and not a destination. Excellence in STEM education should not only be associated with high performance in international achievement tests such as TIMSS and PISA. Even though achieving top performance in these tests has always been likened to obtaining medals in the "Olympics" of education (Leung, 2014), this is a very narrow view of what it means to be successful in STEM education. Just as in the Philippines and Thailand, declining performance over the years in these achievement tests has triggered calls in various countries to reform education policies, professional learning, and practices. While this initiation to reform is a good first step, it is important to acknowledge that improving education standards, and STEM education, in particular, requires us to get a myriad of educational components operating together in specific contexts for different countries.

Here, I use the metaphor of confluences – where two or more rivers, each with their own flow and paths, meet to form a bigger river – to characterise how the Philippines and Thailand improve STEM teacher education in their respective contexts. Following Choy (2021), I view the whole enterprise of improving STEM teacher education as the coming together or flowing together of different educational aspects for a single purpose: To provide all our students with quality STEM learning experiences so that they are supported to achieve the desired learning outcomes. There are two aspects of confluences here. First, there is a directed flow of policies, professional learning initiatives, and practices towards the same goal of providing high-quality STEM learning experiences for all. Second, there is a coming together of different understandings about the main elements of an excellent STEM education. The idea is not to have a single understanding about what or how to teach because understandings and perceptions of STEM education, integrated STEM, and STEM pedagogies are mixed and not clearly defined (English, 2016; Margot & Kettler, 2019). Rather, the aim is to achieve a *balance* point in which our different understandings about STEM teaching and learning are *compatible*. For example, a problem-centric approach to STEM integration (Tan et al., 2019) may not be the only way to ensure the integrated nature of STEM learning activities – one could adopt a solution-centric or a user-centric approach to STEM integration (Teo et al., 2021) – and adopting different centricities of STEM integration may be appropriate at times.

According to Choy (2021), the idea of confluences suggests that improving STEM teacher education is a journey specific to the country. It highlights that each country will need to take its own path to achieve excellence. What is needed then is not to transplant policies, professional learning initiatives, and classroom practices from one country to another. Instead, efforts should be expended to ensure that the policies, initiatives, and classroom practices are all in sync with the purpose of providing high-quality STEM learning experiences, as envisioned by the country. While it is useful to take reference from extant literature on STEM education to have an idea of what quality STEM education may look like, each country has to find its own balance point and get the policies, initiatives, and practices to "flow" in sync.

Seeing the journey of the Philippines and Thailand as confluences of policy, professional learning, and practices provides the opportunity for us to examine the coherence of STEM education implementation in a more holistic way. As described in Teo and Choy (in-press), although Singapore has no official integrated STEM curriculum, there are concerted efforts to formulate educational policies and provide supportive educational environments for the purpose of providing quality STEM experiences to all learners. This is so despite the fact that the Ministry of Education in Singapore, the Singapore Science Centre, the meriSTEM@NIE (a research centre at the National Institute of Education), specialised schools in STEM fields, and schools in Singapore have different views about STEM integration. Yet, the different pathways and ideas adopted by the various stakeholders are built on the strong foundations of the Singapore education system to achieve the common goal of the desired outcomes of education. These initiatives and ideas meet the same purpose. Likewise, the journeys taken by the Philippines and Thailand suggest that achieving quality STEM teacher education lies at the confluence of policies, professional learning, and practices related to STEM education. The two countries "harmonise" their approaches to improve STEM teacher education according to what matters and what works in their respective contexts. Although not every single policy, initiative, and practice may succeed at first, it is important that these policies, initiatives and practices are moving towards a common vision for STEM education. Each country has to find their own paths based on their own strengths and contexts for their journey towards achieving a quality STEM education for their learners through the enhancement of their teacher education.

References

Anderson, J., & Tully, D. (2020). Designing and Evaluating an Integrated STEM Professional Development Program for Secondary and Primary School Teachers in Australia. In *Integrated Approaches to STEM Education* (pp. 403–425). https://doi.org/10.1007/978-3-030-52229-2_22.

Bybee, R. W. (2010, Sep 2010 2017-11-17). Advancing STEM Education: A 2020 Vision. *Technology and Engineering Teacher*, 70(1), 30–35. http://libservy.nie.edu.sg/

login.php?url=https://www.proquest.com/scholarly-journals/advancing-stem-education-2020-vision/docview/853062675/se-2?accountid=28158

Choy, B. H. (2021). Excellence in mathematics education: Multiple confluences. In Y. H. Leong, B. Kaur, B. H. Choy, B. W. J. Yeo, & S. L. Chin (Eds.), *Excellence in Mathematics Education: Foundations and Pathways (Proceedings of the 43rd annual conference of the Mathematics Education Research Group of Australasia)* (pp. 53–56). Singapore: MERGA.

Desimone, L. M. (2009). Improving Impact Studies of Teachers' Professional Development: Toward Better Conceptualizations and Measures. *Educational Researcher, 38*(3), 181–199. https://doi.org/10.3102/0013189x08331140.

English, L. D. (2016). STEM Education K-12: Perspectives on Integration. *International Journal of STEM Education, 3*(1). https://doi.org/10.1186/s40594-016-0036-1.

Fullan, M. (2006). Change theory: A force for school improvement (Centre for Strategic Education Seminar Series Paper 157). http://michaelfullan.ca/wp-content/uploads/2016/06/13396072630.pdf

Hattie, J. (2003). Teachers make a difference: What is the research evidence? Australian Council for Educational Research Annual Conference, Melbourne, Australia.

Lertdechapat, K. (2020). The development of teachers' pedagogical content knowledge for STEM teaching through lesson study to enhance students' 21st century learning and innovation skills [Doctoral dissertation]. Kasetsart University.

Leung, F. K. S. (2014). What Can and Should We Learn from International Studies of Mathematics Achievement? *Mathematics Education Research Journal, 26*(3), 579–605. https://doi.org/10.1007/s13394-013-0109-0.

Lewis, C., Friedkin, S., Baker, E., & Perry, R. (2011). Learning from the Key Tasks of Lesson Study. In O. Zaslavsky, & P. Sullivan (Eds.), *Constructing Knowledge for Teaching Secondary Mathematics* (pp. 161–176). Springer: New York, USA. https://doi.org/10.1007/978-0-387-09812-8_10.

Li, Y., & Anderson, J. (2020). Developing teachers, teaching, and teacher education for integrated STEM education. In *Integrated Approaches to STEM Education* (pp. 353–360). https://doi.org/10.1007/978-3-030-52229-2_19.

Margot, K. C., & Kettler, T. (2019). Teachers' Perception of STEM Integration and Education: A Systematic Literature Review. *International Journal of STEM Education, 6*(1). https://doi.org/10.1186/s40594-018-0151-2.

Rinke, C. R., Gladstone-Brown, W., Kinlaw, C. R., & Cappiello, J. (2016). Characterizing STEM Teacher Education: Affordances and Constraints of Explicit STEM Preparation for Elementary Teachers. *School Science and Mathematics, 116*(6), 300–309. https://doi.org/https://doi.org/10.1111/ssm.12185.

Shulman, L. S. (1987). Knowledge and Teaching: Foundations of the New Reform. *Harvard Educational Review, 57*(1), 1–22.

Tan, A. L., Teo, T. W., Choy, B. H., & Ong, Y. S. (2019). The S-T-E-M Quartet. *Innovation and Education, 1*(1). https://doi.org/10.1186/s42862-019-0005-x.

Teo, T. W., & Choy, B. H. (in-press). STEM Education in Singapore. In O. S. Tan, E. L. Low, E. G. Tay, & Y. K. Yan (Eds.), *Singapore Math and Science Education Innovation: Beyond PISA.* Springer.

Teo, T. W., & Ke, K. J. (2014). Challenges in STEM Teaching: Implication for Preservice and Inservice Teacher Education Program. *Theory Into Practice, 53*(1), 18–24. https://doi.org/10.1080/00405841.2014.862116.

Teo, T. W., Tan, A. L., Ong, Y. S., & Choy, B. H. (2021). Centricities of STEM curriculum frameworks: Variations of the S-T-E-M quartet. Manuscript submitted for publication.

4 STEM teacher education in the Philippines

Gladys C. Nivera, Auxencia A. Limjap, Edwehna Elinore S. Paderna, and Crist John M. Pastor

Defining "STEM education" and "STEM teacher education"

Despite the tremendous attention that Science, Technology, Engineering, and Mathematics (STEM) education receives worldwide, there is no consensus on what it constitutes and what it means in terms of curriculum, pedagogy, and student outcomes (Holmlund et al., 2018; Lamberg & Trzynadlowski, 2015). However, an increasingly common view of STEM education is an integrated, interdisciplinary, and applied approach to learning that brings all four disciplines – Science, Technology, Engineering, and Mathematics, together (Southwest Regional STEM Network, 2009). A variation of STEM is STEAM, which includes an "A" for art and design or agriculture. In the Philippines, the term "STEM education" may be understood as the Mathematics and Science subjects offered in the basic education curriculum or as the strand in the senior high school programme that prepares students for degrees in Science, Technology, Engineering, or Mathematics. Science and Mathematics have always been taught separately in basic education. An integrated and applied approach to teaching these four disciplines is neither evident in the STEM strand curriculum nor observed in actual practice. Occasionally, select Science-oriented schools may offer courses such as robotics, coding, and engineering design thinking, which involve integrating the STEM disciplines. At best, in the general curriculum, the integration of concepts and skills in STEM occurs mostly in capstone projects or performance tasks.

Despite this limited conceptualisation of STEM education in the country, the influence of global practices and literature has led some local educators, researchers, and politicians to use the term STEM as a descriptive term for a "heuristic approach to provide desired 21st-century skills and competencies" (Teng, 2019) and to define STEM skill sets as "those needed to produce more scientists and engineers to power economic growth" (Teng, 2019). In fact, in this era of knowledge-based society and knowledge economy, STEM education is widely seen among emerging economies like the Philippines as a roadmap to innovation and economic progress (Padolina, 2014). Within these contexts, this paper uses the term *STEM teacher education* as programmes designed to

DOI: 10.4324/9781003099888-4

develop future Mathematics and Science teachers for basic education. A *STEM teacher* refers to a Mathematics or Science teacher in basic education. In contrast, a *STEM teacher educator* refers to a faculty in a teacher education programme teaching Mathematics or Science content or pedagogical content knowledge (PCK) courses to future Mathematics and Science teachers.

STEM education in the Philippines: Issues and concerns

In 2013, the Philippine government embarked on an ambitious education sector reform programme to increase national competitiveness (World Bank, 2016). Amidst growing concerns that the ten-year basic education curriculum was constraining national competitiveness, the Philippines passed the 2013 Basic Education Act, more commonly known as the K to 12 programme. This Act, which aligned the Philippines' educational system with its global counterparts, extended the basic education cycle from 10 to 13 years by including Kindergarten and two years of senior high school. Senior high school students in the academic track may go to any one of these strands: STEM, Humanities and Social Sciences (HUMSS), Accountancy and Business Management (ABM), and General Academic Strand (GAS).

With the roll-out of the STEM strand in senior high school in 2016, STEM became a byword for the pursuit of challenging but more rewarding and prestigious careers for college-bound students. This newfound status and prominence of STEM education came against a backdrop of unpleasant Mathematics and Science education realities. The Philippines ranked 78th out of 79 countries in mathematics and scientific literacy in the *Programme for International Student Assessment* (PISA, 2018). Likewise, it ranked last among 58 countries in Mathematics and Science assessment for Grade 4 students in the *Trends in International Mathematics and Science Survey* (TIMSS, 2019). Sadly, the country has yet to see a significant improvement in its mathematics and science performance in international metrics despite the K to 12 reform in 2012.

In its 2019 Year-End Report, DepEd described the Filipino students' performance in the National Achievement Test (NAT), an annual large-scale assessment it conducts for Grade 6, 10, and 12 students, as "gravitating towards the low proficiency levels", especially in Science, Mathematics, and English (Gonzales, 2019). It viewed the country's poor performance in these international and national assessments as an urgent call to address issues and gaps in the educational system and implement aggressive reforms (DepEd, 2019a). In December 2019, DepEd launched *Sulong Edukalidad*, which is its rallying call for a national effort for quality education. *Sulong Edukalidad*'s four pillars of aggressive reforms for quality are: (1) K to 12 curriculum review and update; (2) Improving the learning environment; (3) Teachers' upskilling and reskilling; and (4) Engagement of stakeholders for support and collaboration (DepEd 2019b).

Arguably, quality learning is contingent upon quality teaching (DepEd, 2017) and that a teacher's knowledge influences their students' Mathematics achievement (Tatto et al., 2012). Given the country's poor performances in TIMSS,

PISA, and the NAT, it is perhaps not surprising to find that its elementary and high school teachers' knowledge of subject matter in Mathematics and Science is low. In the survey conducted by the World Bank with some assistance from the DepEd and the Philippine Normal University's (PNU) Research Centre for Teacher Quality, the median elementary Mathematics teacher was able to answer just a little over 40% of the questions on Grade 6 Mathematics content. The median elementary Science teacher answered a little less than 20% of the Grade 6 Science content questions in comparison. The high school Mathematics and Science teachers who were specialists in their disciplines did not fare any better. The median Mathematics and Science teachers answered only about 30% of the Grade 10 Mathematics and Science content questions. The report suggested that for teachers to implement the curriculum and provide effective instruction in the classroom, the teacher's skills and subject knowledge have to improve significantly (World Bank, 2016).

Apart from teacher quality, the basic education sector has to deal with the shortage of subject specialist teachers. With the introduction of the senior high school in 2016, the demand for Mathematics and Science teachers far outstrips the supply, despite DOST-SEI's aggressive scholarship programmes in Mathematics and Science Education. Estimates showed that even if all licensed teachers were hired, they would still not be enough to fill the need for Mathematics and Science teachers (David & Ducanes, 2018).

In the World Bank Philippines Public Expenditure Tracking and Quantitative Service Delivery Study, the school principals interviewed reported a shortage of Mathematics and Science teachers equivalent to about 30% of the workforce (World Bank, 2016). The number of Filipino teachers working abroad contributed to this shortage of teachers. According to the 2019 survey on overseas Filipinos by the Philippine Statistics Authority (2020), in the years 2013 to 2017, the Philippines deployed an average of 1500 teachers worldwide every year. Predictably, the teachers who qualify and choose to work abroad are the competent ones. These teachers cite meagre salaries, difficult work environment, and lack of professional growth among the reasons they leave the Philippines (Novio, 2019). In 2018, the Department of Labor and Employment (DOLE), in cooperation with other agencies, implemented the "Sa 'Pinas, Ikaw ang Ma'am at Sir" programme to encourage teachers deployed overseas to practise their profession in the Philippines (Aurelio, 2017). Within the year of its inception, the programme provided a total of 1,421 positions for returning licensed teachers (DOLE, 2018).

While the quality and quantity of teachers are important determinant factors of student learning outcomes in the Philippines, these are not the only critical ones. The 2016 World Bank assessment of the Philippine basic education service delivery showed that the country spent less per student per capita gross domestic product (GDP) share than most middle-income countries. Based on purchasing power parity conversions, the Philippines only spent US$380 per elementary student compared to US$760 in Vietnam and US$2350 in Malaysia (World Bank, 2016). DepEd gets the biggest slice of the national budget at 3.4% of the GDP as mandated by the constitution (Teodoro, 2020). However, it is less than that of Vietnam with 6.3%, Malaysia with 6.1%, or even Laos with 4.2% (Medenilla, 2019). Undoubtedly, the country needs to raise its investment in education significantly to increase access to quality education.

The low investment in education has spawned various problems for the basic education sector, such as classroom shortages, poor and insufficient facilities, laboratory equipment, and instructional materials, low access to quality education, and high attrition rates among students from low socio-economic backgrounds (Ambag, 2018; Cristobal, 2019; John & Estonato, 2017; Lee et al., 2019; Nivera, 2018; World Bank, 2016). A STEM education pipeline study in the Philippines showed that in one cohort, 18% of the Grade 1 pupils did not make it to Grade 2, and only 7% of the Grade 1 pupils eventually completed a college degree in STEM (Nivera, 2018). The high attrition rate and the small number of STEM degree graduates produced by the pipeline keep the country from achieving its goal to have a strong and steady supply of STEM professionals to support national economic progress. Studies show that the following factors motivate Filipino students to pursue STEM careers: Having competent STEM teachers, exposure to various hands-on and investigative experiences in STEM classrooms, engaging in extracurricular activities such as Math/Science camps and quiz bees, access to scholarships, and awareness or exposure to STEM careers and industries (Bernardo et al., 2008; Nivera, 2018).

Undoubtedly, quality STEM education makes a difference in ensuring a sufficient and continuous supply of STEM professionals. Since a robust STEM education is becoming more important to our increasingly complex world, the Philippines must increase its investment in STEM education to develop and continuously grow its pool of competent STEM teachers.

Governance of teacher education programmes

To decongest a huge bureaucracy and ensure individual and more focused attention to the different levels of education, the government split up the Department of Education, Culture, and Sports into the DepEd, the Commission on Higher Education (CHED), and the Technical Education and Skills Development Authority (TESDA) in 1994. DepEd was to supervise basic education, TESDA to manage the technical-vocational and middle-level education, and CHED to supervise higher education. This move unintentionally resulted in the fragmentation of the goals, plans, and projects in education. The Teacher Education Council (TEC) was created in the same year to serve as an umbrella organisation where the various agencies could harmonise the plans, programmes, and projects for teacher education. Its primary mandate was to formulate policies to strengthen and improve teacher education quality and designate Centres of Excellence (COE) and Centres of Development (COD) for teacher education in partnership with CHED. The designation as a centre carries certain privileges, particularly in terms of priority funding for government projects. In return, the designated centre has to take a leadership role in their province or region and assist and serve other teacher education institutions (TEIs), as needed. At present, there are 74 COEs and CODs in teacher education out of the 1572 TEIs in the country (OSWG, 2020). While some universities are also identified by CHED as COEs or CODs in Science and Mathematics, in Engineering, and in Information Technology Education, none is identified for STEM or STEM education.

CHED's mandate is to promote relevant and quality higher education, protect academic freedom, monitor and evaluate the performance of programmes and institutions of higher learning, and ensure the accessibility of higher education to all. It sets the minimum standards for programmes of institutions of higher learning recommended by panels of experts in the field and subject to a public hearing. Based on CHED records, teacher education is now the most heavily subscribed college programme. For the academic year 2019–2020, the country had a total of 1975 higher education institutions (excluding satellite campuses), 1572 of which have teacher education offering one or both of the two major teacher education degrees authorised by CHED – the Bachelor in Secondary Education (BSEd) and the Bachelor in Elementary Education (BEEd). Including satellite campuses, the Philippines has 2396 higher education institutions, of which 246 are public universities and colleges, and 1729 are private institutions (CHED, 2020).

Although the number of TEIs spread all over the country improves access for aspiring teachers, it also creates problems in assuring quality and high standards. The performances of TEIs in the Board for Licensure Examination for Professional Teachers (BLEPT) vary greatly. The Philippine Business for Education (PBEd) reported that while top-tier TEIs enjoy close to 100% passing rates for their graduates, at least half of the TEIs perform below the national passing rate, and others fail to produce a single BLEPT passer (Reyes, 2019). Demanding the closure of non-performing programmes is difficult as it may require legal action and involve expensive litigation.

Quality assurance policies and arrangements can make a difference in teacher education (World Bank, 2016). Policies must be in place to make teaching an attractive career, assure the quality of entrants to teacher education, promote and support the accreditation of teacher education programmes, and ensure that graduates have high-performance standards before entering the teaching profession. CHED ensures compliance with the policies, standards, and guidelines for teacher education programmes by requiring institutions to obtain a Certificate of Programme Compliance for its offerings. While strongly encouraged by CHED, programme or institutional accreditation of higher education institutions by private organisations is voluntary on the part of the institution.

It is important to note that the Philippine Professional Standards for Teachers (PPST) now underpins most of the current reforms and quality assurance efforts of the DepEd, CHED, TEC, and the Professional Regulation Commission (PRC). The PPST is a document that "defines teacher quality in the Philippines" and outlines the needed competencies and skills of quality teachers to enable them to manage and handle emerging global frameworks (DepEd, 2017). The standards describe the expectations of teachers' increasing levels of knowledge, practice, and professional engagement. According to the DepEd order, the PPST shall be used as the basis for all learning and development programmes for teachers to ensure that teachers have what it takes to implement the K to 12 programme effectively. Further, the PPST aims to: (1) Set clear expectations of teachers along well-defined career stages of professional development from beginning to distinguished practice; (2) Engage teachers to embrace a continuing effort in attaining proficiency actively; and (3) Apply a uniform measure to assess teacher performance, identify needs, and provide support for professional development (DepEd, 2017). PPST impacts all areas of teacher education. However, "PPST only includes general attributes of teacher quality, proficiency, and career stages but with no elaborations on subject matter or content – as well as teaching and learning of complex skills in the tertiary level" (Morales et al., 2019). To complement the PPST, a standard self-rating tool to determine the Philippine Higher Education (PHE) STEAM (Science, Technology, Engineering, Agri-Fisheries, and Mathematics) Educators' proficiencies was developed for reflective practice and policy inputs to PHE-STEAM (Morales et al., 2019). Whether this tool for rating the STEAM educators' proficiencies will influence the STEM teacher education programmes and the professional development of STEM education teachers in the country in any significant way remains to be seen pending CHED's adoption or endorsement of the said tool.

The pathway of STEM teachers

Entry into teacher education programmes

Arguably, the quality of future teachers depends to some extent on the quality of entrants to the teacher education programme. For many decades, teacher education has not been a popular choice by top high school graduates and male students in the Philippines as teaching as a profession is not considered by many as "high paying" and "prestigious". The teacher education programme's easy accessibility and low entry requirements make it a convenient choice for the less privileged and those whose life options are limited (Bilbao & Nivera, 2014). However, raising the entry requirements to assure the quality of entrants to teacher education on a national level is difficult and contentious, particularly for private institutions as such will effectively decrease the number of enrollees. On the upside, the teaching profession's image is slowly improving due to the improved employment opportunities for teachers and yearly salary increases from 2016 to 2023 under the Salary Standardisation Laws (SSL) of 2016 and 2020. In 2015, the entry-level teacher I's monthly salary was PhP 19,218, or about US $392. By 2023, it will be PhP 27,000, or about US $551, an increase of about 40% in 8 years' time (Department of Budget and Management, 2015; SSL, 2019). Graduates of teacher education are almost guaranteed to get good and secure jobs, which is a strong motivating factor to go into teaching. Top high school students are now beginning to find teaching as an attractive and secure option.

The DOST-SEI and other public agencies and private companies offer scholarships to top high school graduates to attract quality entrants to the Science and Mathematics teacher education programmes. Likewise, some TEIs conduct career promotion seminars among graduating students of Science-oriented high schools to entice them into Mathematics and Science teaching. While these efforts significantly increase the number of top high school graduates going into STEM teacher education, more still needs to be done to supply the country's need for competent STEM teachers.

Teacher education programmes

Generally, a teacher education programme in the Philippines require four years to complete. The BEEd produces generalists while the BSEd produces specialists in a particular discipline. In 2017, CHED released the current policies, standards, and guidelines for teacher education, which set the minimum required number of units and courses and the core competencies expected of the graduates. While TEIs have the academic freedom to choose and design their curricular offering to fit their mission and context, they are expected to comply with the minimum requirements for specific academic programmes, the General Education distribution, and the specific professional courses. The CHED teacher education curricula claim to reflect the shift to outcomes-based education and respond to the K to 12 reform and the Philippine Qualifications Framework, among other drivers. Moreover, they claim to address the requisites of Education 4.0, which, according to the World Economic Forum (2020), is characterised by the provision of students with skills for global citizenship, innovation and creativity, technology, interpersonal, personalised, and self-paced learning, accessible and inclusive learning, and problem-based and collaborative learning. Graduates of CHED's teacher education curriculum are expected to be abreast with technological advances and apply these technology tools for instruction and innovate, design, and engineer materials and technologies to gather data for classroom use.

Unfortunately, the allocation of courses in the teacher education curriculum does not reflect this need to emphasise and maximise the potential of technology in instruction. For instance, technology courses are allotted only three units in BEEd and six units in BSEd. Not only are these courses limited in number, but they also only teach teachers to be end-users of technology for instruction rather than users of technology for innovation and design thinking. The teacher education curriculum may consider more innovative, non-traditional technology-enhanced courses that develop computational thinking, design thinking, transferrable competencies, and the like. Understandably, such innovations will require a corresponding, aggressive, and focused capacity-building efforts for the TEI's faculty, and the upgrading of equipment and infrastructure, which many institutions will find very challenging.

A recent local study explored the TEI's readiness for Education 4.0 (Alda et al., 2020). In this study, the administrators and faculty members believed that they were ready to integrate digital resources for teaching and learning. However, they also believed that they were neither skilful in using the existing learning management system and other online class modalities nor in augmented reality, robotics, and other digital enablers, especially since they were not available in most institutions (Alda et al., 2020). To be future-ready, TEIs have to rethink and redesign their programmes, equip their faculty in breakthrough areas that promote new ideas and ways of thinking, and support such reforms with the needed infrastructure.

Eligibility to the teaching profession

To assure the quality of graduates that will enter the teaching profession, the PRC conducts the Board Licensure for Professional Teachers (BLEPT) twice a year. The multiple-choice written examination for future high school teachers covers content in the General Education, Professional Education, and Specialisation courses. Future elementary teachers take a similar examination but without the specialisation component. A BLEPT passer becomes a licensed professional teacher eligible to work in basic education.

The BLEPT is the lowest-performing licensure examination in the country, with more than half of the TEIs performing below the national average of 50% (Torre, 2017). Even with rigorous review programmes conducted by TEIs and other private review centres to prepare test takers for the BLEPT, the average passing rate from 2010 to 2017 was only about 35% for secondary teachers and about 30% for elementary teachers (David & Ducanes, 2018). These low passing rates may reflect on the quality of training they received from the TEIs, the entrants' qualifications, or the validity of the BLEPT. Despite numerous calls for the BLEPT items to be subjected to the scrutiny of experts and validity testing, these items have never been made public (Torre, 2017). At present, the BLEPT is undergoing review and revision in light of the K to 12 reforms in basic education, the shift to outcomes-based education, the new policies and standards in higher education, and DepEd and CHED's adoption of the PPST.

In-service training

An In-Service Training for Teachers (INSET) is one of the tools towards quality education. While teachers in both basic and higher education regularly attend one to five-day INSETs focused on their personal and professional developments, intensive training programmes to upgrade the teachers' competencies and PCK in Mathematics and Science are often offered only to a select few. Upon completing such training programmes, the teacher-participants usually share their learnings with their co-teachers through an "echo" seminar. The cascade training model is viewed as an efficient and cost-effective way of "disseminating knowledge" in an exponential manner. Karalis (2016) suggests that the cascade approach is the best choice for knowledge transfer and when dealing with a large number of final recipients in parallel with limited time and cost. However, he cautions that this mechanism works best with continuous monitoring of the programme's unfolding and well-prepared and complete educational materials. Unfortunately, most cascade training programmes in the country do not have these two requirements. Moreover, questions remain on the effectiveness of the cascade model in education in changing behaviours, attitudes, perceptions, and equipping participants with new skills, which are all beyond "knowledge transfer".

INSET evaluation as part of quality assurance often comes in a simple survey form filled out by participants and rarely done through actual monitoring and research. As World Bank (2016) observed, DepEd has many programmes whose effectiveness has not been adequately evaluated before and after implementation, making it difficult to keep track of the actual "value-added" by these interventions. Another concern is the one-size-fits-all approach in designing the INSET in the absence of a timely and appropriate national teachers' training needs assessment. A study of the INSETS conducted for SY 2019–2020 showed that almost all of them were content-based with limited hands-on activities and application in the teachers' actual classroom practices, and very limited integration of Education 4.0 (Tupas & Noderama, 2020). An exception to this trend is DOST-SEI's Science Teacher Academy for the Regions (STAR), a cluster of innovative capacity building activities aimed to improve the quality of teaching of STEM teachers. In 2021, STAR conducted a series of online training workshops on Design Thinking among K – 3 Science and Mathematics teachers in various regions (DOST-SEI, 2021).

Occasionally, private companies partner with higher education institutions and a government agency to launch specialised and innovative training programmes for select teachers. The Unilab Foundation, University of the Philippines College of Education, and DOST-SEI co-sponsored and co-organised the series of training programmes on *The Engineering Design Process*, which aimed to build and strengthen the capability of select STEM teachers by relating 21st-century competencies and literacies to STEM learning. The participants identified a specific community or industry problem in their locality and generated possible solutions to the problem using the design thinking process. However, these innovative and community-based training programmes are open only to a very limited number of teachers and schools.

To promote teacher quality, the National Educators' Academy of the Philippines (NEAP), with the assistance of the PNU Research Centre for Teacher Quality, is currently designing a system for effective implementation of professional development programmes for teachers anchored on the PPST. One such ongoing collaborative project between DepEd-NEAP and PNU is the Linking Standards and Quality Practice (LiSQuP), which targets 2820 DepEd teachers, school leaders and education personnel to take up a Masters, doctorate, or executive program at PNU. LiSQuP consists of academic coursework, job-embedded learning activities and practice, and specialised and focused learning (RCTQ, 2021).

Degree-earning professional development programs are expected to have more lasting impact towards quality education than the short-term INSETs.

Mathematics and Science teacher education programmes

In this discussion, STEM teacher education programmes refer to those that develop the Mathematics and Science teachers for secondary education in the Philippines, namely, the BSEd major in Science and BSEd major in Mathematics. Both programmes have four components, namely, General Education (36 units), Professional Education (48 units), Mandated (6 units), and Specialisation (73 units for the BSEd major in Science and 63 units for the BSEd major in Mathematics). A typical course per semester has 3 units requiring 54 hours of contact time. Science courses with lecture and laboratory components often have 4 units, that is, 3 units (54 hours) for the lecture and 1 unit (54 hours) for the laboratory. It is only in laboratory courses where 1 unit is equivalent to 54 hours. The Professional Education includes 6 units of field study, which immerse future teachers in actual classroom situations in the field, and 6 units of practice teaching in private or public basic education schools.

Bachelor in Secondary Education (BSEd) major in Mathematics programme

The four-year BSEd major in Mathematics programme requires a minimum of 155 units. Its 63-unit specialisation courses consist of 51 units of Mathematics content courses, 9 units of PCK courses focused on strategies, instrumentation, and assessment, and 3 units of research in Mathematics education. Shulman (1987) argued that PCK courses are critical components of teacher preparation as they blend content and pedagogy to enhance the understanding of how particular topics, problems, or issues are organised, represented, and adapted to the diverse interests and various levels of learners' abilities. A study on the Mathematics programmes of some TEIs in the Philippines argued that the small number of PCK courses in the BSEd major in Mathematics curriculum did not provide the right balance between content and PCK (Nivera et al., 2017). While CHED only sets the curriculum's minimum requirements and institutions are free to add more units, few institutions do, and those who do tend to add more content courses rather than PCK courses (Nivera et al., 2017).

The Teacher Education and Development Study in Mathematics (TEDS-M), the first cross-national study on Mathematics teacher education with a large-scale sample, claims that the design of the teacher education curricula can have substantial effects on the level of knowledge that future teachers can acquire (Tatto et al., 2012). The report showed a weak but consistent association between teachers' pattern of beliefs and their performance on the knowledge in the Mathematics test. The countries most strongly endorsing the beliefs consistent with the conceptual orientation (meaning that Mathematics is a process of inquiry and learning it requires active involvement) were generally those with higher mean scores on the knowledge test. The countries most strongly endorsing the beliefs consistent with the calculation orientation (meaning that Mathematics is a set of rules and procedures, a fixed ability, and learning it requires teacher direction) were generally those with lower mean scores on the knowledge test. The study further reported that, without exception, the pattern of beliefs held by the future teachers in every country matched the patterns of beliefs held by the teacher educators (Tatto et al., 2012). These findings suggest that beyond altering the teacher preparation curriculum and providing more PCK courses, TEIs may do well by looking into the beliefs held by their Mathematics teacher educators.

Bachelor in Secondary Education (BSEd) major in Science programme

Traditionally, the teacher education programme for Science teachers at the secondary level included a specialisation in Physics, Chemistry, Biology, or General Science. Such design catered well to the compartmentalised Science curriculum in secondary schools where Year 1 focused on General Science, Year 2 on Biology, Year 3 on Chemistry, and Year 4 on Physics. The shift to the spiral organisation of the Science content in junior high school in 2012 forced all Science teachers to teach the four specialisations at every year level. As expected, the Biology teachers struggled to teach Chemistry and Physics, while the Physics teachers struggled to teach Biology and Chemistry. To add to their woes, they had to teach the various areas of Science in an integrated manner, which was a stark departure from how the TEIs trained them.

In response to this mismatch between teacher preparation and demands in the field, CHED released a new set of policies, standards, and guidelines for the BSEd major in Science programme in 2017 to adequately prepare Science teachers to teach the spiral Science curriculum in basic education. Its graduates are expected to demonstrate a deep understanding of scientific concepts and principles, apply scientific inquiry in teaching and learning, and utilise effective Science teaching and assessment methods.

This revised 4-year programme has 165 units in all, with 73 units allotted for major courses split into content (67 units) and PCK (6 units). Its graduates are expected to be competent in Biology, Chemistry, Environmental, Earth and Space Science, and Physics, but a specialist in no particular area. The PCK courses make up a very small percentage (8%) of the major courses, which reinforces the observation that the development of PCK does not receive the importance that it deserves in pre-service teacher preparation in the Philippines (Nivera et al., 2017).

Further, teaching the Sciences in an integrated manner requires knowledge of the separate disciplines in Science and a broad understanding of their interconnectedness. A three-unit course in the curriculum guides teachers in teaching integrated Science. However, there are concerns that a three-unit course may not be sufficient for pre-service teachers to internalise the synergy and interaction among the Sciences and to reflect these in the pedagogy, instructional materials, and assessment practices in their classrooms.

Innovations in the Mathematics and Science teacher education programmes

In this section, we will present some innovations in select Mathematics and Science teacher education programmes. These innovative curricular offerings and courses are exceptions as TEIs in the Philippines generally follow the CHED suggested curricula.

Bachelor in Science education with specialisation in Biology/Chemistry/Physics

By its status as the National Centre for Teacher Education, PNU has the freedom to implement innovative Mathematics and Science programmes for training and research purposes. Its teacher education programme for Science teachers contrasts with CHED's BSEd major in science by allowing students to specialise in Biology, Chemistry, or Physics while preparing them to teach the spiral Science curriculum in junior high school. Thus, PNU's science education graduates can teach the integrated Science courses in junior high school and the compartmentalised Science courses (e.g., Physics, Chemistry, and Biology) in Science-oriented junior high schools and senior high schools. The specialisation also allows them to pursue a master's degree in Biology, Chemistry, Physics, or Science Education. Table 4.1 compares the number of units of the CHED's BSEd major in science programme and PNU's Bachelor in Science Education (BSciE) programme with a specialisation in Biology, Chemistry, or Physics.

PNU's BSciE programme includes six more units of PCK courses than that suggested by CHED to deepen the future teachers' understanding of the concepts and theories of learning, curriculum, pedagogy, and assessment in Science

Table 4.1 Comparison of the BSEd major in Science and BSciE Bio/Chem/Phy

Courses	Number of units in CHED's BSEd major in Science	Number of units in PNU's BSciE major in Biology/Chemistry/Physics
General Education	36	36
Professional Education	42	45
Specialisation (core and major)	73	76–77
Pedagogical content knowledge	6[a]	12
Electives	0	6
Mandated courses (PE and NSTP)	14	14
Total for programme	165	189–190

[a]Included in the specialisation subjects.

education. Another unique feature is the three-unit Summer Science Internship Course, which provides students with first-hand industry and laboratory experiences and fieldwork to enhance their understanding of the applications of scientific concepts and skills in the real world, and build networks with experts and actual scientists in the field. Students take their Summer Science Internships in government agencies such as the Philippine National Research Institute, the National Museum, and the Philippine Institute of Volcanology and Seismology. This programme is the only one in the country that offers a pure Science internship on top of the teaching internship in basic education schools.

Bachelor in Mathematics education

PNU's 178-unit Bachelor in Mathematics Education programme goes beyond CHED's required 165 units for BSEd major in Mathematics. The additional units were devoted to PCK courses and elective content courses to provide students with a more holistic and broader preparation. For instance, it offers an innovative elective course titled "English for Mathematics" because of the Mathematics teachers' perceived weakness in mathematical reasoning and communication. Other elective courses include the use of LATeX (software system) and other programming languages and Physics.

Bachelor in Mathematics and Science for elementary education

The Bachelor in Mathematics and Science for Elementary Education (BMSEE) aims to address the criticism that the typical elementary teachers who are "generalists" do not have the necessary competence to teach Mathematics and Science with the desired rigour and depth. BMSEE offers 16 additional units in Mathematics and Science to the CHED-mandated "generalist" elementary teacher education programme. Other specialisation courses were fused or integrated to accommodate the additional content and PCK courses without overpacking the curriculum. Table 4.2 gives a quick comparison of the number of Mathematics and Science units in the CHED's BEEd and PNU's BMSEE programmes.

As a programme that develops elementary teachers with specialisations in Mathematics and Science, BMSEE is the first of its kind in the country, but the idea is not new. In countries like Japan, teachers in the primary level have a major

Table 4.2 Comparison of the number of units in BEEd and BMSEE

Courses	Number of units in CHED's BEEd	Number of units in PNU's BMSEE
Math content courses	6	12
Science content courses	6	16
PCK for Math and Science	0	12
Total for programme	152	177

in a Science-related course, and most of them are pursuing graduate degrees in their specialisation (Penerio & Toshihiko, 2020). Likewise, the TEDS-M report provides a strong argument for preparing Mathematics specialists to teach elementary Mathematics. The study showed that the average, future primary teachers prepared as Mathematics specialists had higher Mathematics content knowledge and Mathematics PCK scores than those prepared to teach as lower-primary generalists (Tatto et al. 2012). Countries like Germany, Malaysia, Singapore, and Thailand prepare specialist teachers of Mathematics to teach below Grade 6.

Course on teaching STEM research in basic education

In some cases, higher education institutions introduce innovation into their teacher education programmes through particular courses. In 2018, the University of the Philippines instituted a course titled *Teaching STEM Research in Basic Education.* The course aims to equip future Science and Mathematics teachers with the knowledge and relevant techniques and strategies to develop 21st-century skills and mentor them in generating developmentally appropriate innovations. It promotes STEM practices geared towards developing an innovation mindset among learners from diverse contexts and empower them to generate innovative products or optimised processes/services that are developmentally appropriate and may have economic or social value. CHED and TEIs may do well to consider the inclusion of similar innovative STEM courses in their Mathematics and Science Teacher Education curricula to produce future-oriented and innovative STEM teachers who are ready to take on the complex demands of Education 4.0.

Forces of change and further reforms needed

As education worldwide responds to the challenge of the Industrial Revolution 4.0, there is increasing clamour for STEM education to provide students with 21st-century skills (World Economic Forum, 2016). Quality STEM education demands a sufficient supply of competent STEM educators, which puts STEM teacher education at the forefront of any country's efforts to be globally competitive and future ready.

STEM teacher education in the Philippines has developed innovative and competent STEM teachers over the decades who served the Philippines and other countries. As a result of the government's efforts to provide better job opportunities and higher salary grades, the image and status of teaching are slowly improving. STEM teacher education receives significant and multi-sectoral support from the public and private sectors in terms of scholarships, specialised training programmes, research grants, and laboratory equipment and facilities. However, much still need to be done.

The government has to rethink its national policy on education and redirect more of its resources to support this sector. The budget for education as a percentage of the GDP fails in comparison to the percentages allocated for the same sector by its Association of Southeast Asian Nations (ASEAN) neighbours and other middle-income countries. To be competitive, the Philippines needs to raise its investment in education significantly.

STEM teachers are still wanting in quality and quantity due to various political, economic, and societal factors. The quality of STEM teacher education graduates is not assured as the TEIs' standards vary in their entry requirements, programme implementation, available facilities and equipment, and faculty qualifications. The strength of quality assurance arrangements has been shown to have a positive relationship with country performances in Mathematics (Tatto et al., 2012). The countries with weaker arrangements have lower scores than those with stronger arrangements. For instance, in high-performing countries in TIMSS and PISA, such as Singapore and Chinese Taipei, policies are in place to assure the quality of entrants to teacher education (Tatto et al., 2012). These countries can afford to be highly selective of entrants to their teacher education programmes as these programs attract the best and the brightest. Such is not the case in the Philippines.

Despite the low requirements to enter teacher education and the huge number of TEIs spread throughout the country, Philippines still fails to produce the number of STEM teachers required by the system. The low national passing rate in BLEPT (30–35%) means a wastage of about two-thirds of the teacher education graduates. In addition to the measures already mentioned, we need to strengthen supportive teacher policies, provide better working conditions and benefits, and further elevate the teacher's professional and social status to address quality challenges and attract more top graduates of senior high schools to STEM teacher education programmes. Further, the government and the private sector may need to offer more scholarships based on merit and based on needs and cultural background to address questions of access and diversity.

Arguably, different agencies and schools may have an alternative understanding of STEM education. STEM pedagogies such as problem-based learning or integrating Science and Mathematics subjects in the curriculum are not common. Education stakeholders must continue conceptualising STEM implications for the curriculum and attempting innovations based on these different understandings. The issue likewise raises the challenge of curriculum flexibility to allow various innovations.

STEM teacher education programmes may consider offering additional units in PCK and technology utilisation (known as TPCK) to future teachers to ensure understanding of the interweaving of content, theories, and pedagogies for effective teaching and learning. Admittedly, there are tensions between content and pedagogy experts on the issue of priority. However, providing little emphasis on PCK may have a significant and detrimental impact on how teachers teach and, consequently, how students learn. The balance between a content-based curriculum versus a problem-solving-based curriculum may need to be examined as well. Thus, a re-examination of the teacher education curriculum to allow a shift in focus of the curriculum should be considered.

Advancements in Science and technology offer a more interactive and enhanced educational experience for the students in the classroom. While more and more educational institutions are being equipped with facilities for connectivity and mobility in the nation's bid to design collaborative, student-centred learning environments (DOST-SEI, 2018), many of them still have basic facilities only. For the STEM teacher education programmes to offer innovative courses such as STEM Engineering Design Process, STEM Research in Education, Computational Thinking, Coding, or Design Thinking, institutions need to equip their faculty in these new fields and provide the needed equipment and infrastructure.

Further, now that education is no longer space-bound, greater student exchange and mobility should be promoted to develop teachers with a more global perspective and diverse experiences. International programmes like Student Teacher Exchange in Southeast Asia organised by the Southeast Asia Ministers of Education Organisation allow pre-service teachers in Mathematics and Science to do their internship in ASEAN countries. Likewise, the ASEAN Teacher Education Network, the Philippine Teacher Education Network, and the Network of Normal Schools provide student and faculty exchange and research collaboration opportunities among the member institutions. TEIs must also take advantage of the opportunity presented by online classes to invite international lecturers to share their expertise with future teachers.

In this chapter, we have discussed the multiple authorities responsible for education in schools and institutes of higher education. While each authority has well-articulated aims and roles, and while all these authorities include highly professional educators with track records of sound educational decisions, there remains the challenge of liaison between them such that the foci of each are reflected in the processes and policies of the others. In particular, we point to the need for consistency between entry requirements, curriculum standards, instructional strategies, and teacher qualifications. Similarly, if schools are to be given the flexibility to develop integrated STEM programmes and problem-solving approaches, the administrative structures need to be examined as to their flexibility to allowing the implementation of different innovations.

Unfortunately, the multi-sectoral programmes and projects of public and private agencies and organisations geared towards improving STEM education have yet to significantly raise the level of STEM education and STEM teacher education in the Philippines. Simplistic attribution of this low performance to either curriculum, teachers, and teaching is perhaps not productive. Rather, a thorough and extensive whole system critical analysis needs to be done to identify gaps, weaknesses, and appropriate interventions. Finally, a multi-sectoral and data-driven Master Plan for STEM Education need to be considered to harmonise all plans and programmes of various agencies, industry players, and professional and non-government organisations.

While the realities on the ground underscore the complex challenges faced by STEM teacher education in the Philippines, the country is committed to resolving these challenges. The ongoing and massive reforms in both basic and higher education and the multi-sectoral efforts to improve STEM education make us hopeful that the country is firmly moving forward in the right direction.

References

Alda, R., Boholano, H., & Dayagbil, F. (2020). TEIs in the Philippines towards Education 4.0. *International Journal of Learning, Teaching and Educational Research*, *19*(8), 137–154.

Ambag, R. (2018, August 3). *Teaching Science in the Philippines: Why (and How) We Can Do Better.* Flipscience. https://www.flipscience.ph/news/features-news/features/teaching-science-philippines/.

Aurelio, J. (2017, Januray 21). *DOLE Appeals to OFWS to Come Back, Teach in PH.* Inquirer.Net. https://newsinfo.inquirer.net/864152/dole-appeals-to-ofws-to-come-back-teach-in-ph.

Bernardo, A., Limjap A., Prudente, M., & Roleda, L. (2008). Students' perceptions of science classes in the Philippines. *Asia Pacific Education Review*, *9*(3), 285–295. https://files.eric.ed.gov/fulltext/EJ835201.pdf.

Bilbao, P., & Nivera, G. (2014). *The Country Report of the Philippines on Teacher Education.* International Conference on the Teaching Profession in ASEAN, Bangkok, Thailand.

(CHED) Commission on Higher Education. (2020). *Higher Education Statistical Data.* https://ched.gov.ph/wp-content/uploads/Higher-Education-Data-and-Indicators-AY-2009-10-to-AY-2019-20.pdf.

Cristobal, R. (2019, July 20). *Riding into the Future with 'STEM'.* Business Mirror. https://businessmirror.com.ph/2019/07/20/riding-into-the-future-with-stem/.

Department of Budget and Management. (2015). *Salary Standardization Law of 2015.* https://www.dbm.gov.ph/wp-content/uploads/SSL2015/FAQs%20SSL%202015%20as%20of%2011.24.2015.pdf.

David, C., & Ducanes, G. (2018). *Teacher Education in the Philippines: Are We Meeting the Demand for Quantity and Quality?* University of the Philippines Center for Integrative and Development Studies Policy Brief 18-002. https://drive.google.com/drive/folders/16Kel1LfbEkasxzno4Qxa1KIwjmgF96sm.

(DepEd) Department of Education. (2017, August 11). *National Adoption and Implementation of the Philippine Professional Standards for Teachers.* https://www.deped.gov.ph/wp-content/uploads/2017/08/DO_s2017_042-1.pdf.

(DepEd) Department of Education. (2019a, December). *PISA 2018 National Report of the Philippines.* https://www.deped.gov.ph/wp-content/uploads/2019/12/PISA-2018-Philippine-National-Report.pdf.

(DepEd) Department of Education. (2019b, December 3). Sulong Edukalidad: DepEd's Battlecry Moving Forward. https://www.deped.gov.ph/2019/12/03/sulong-edukalidad-depeds-battlecry-moving-forward/.

(DOLE) Department of Labor and Employment. (2018, April 29). *Guidelines Out for OFWs Turned Teachers.* https://www.dole.gov.ph/news/guidelines-out-for-ofws-turned-teachers/.

(DOST-SEI) Department of Science and Technology-Science Education Institute. (2018, March). *Science and Technology Human Resource Development Plan 2017–2022.* http://www.sei.dost.gov.ph/images/downloads/publ/hrdpplan2017-2022.pdf.

Department of Science and Technology-Science Education Institute (RCTQ). (2021, September). *Science Teacher Academy for the Regions (STAR).* https://sei.dost.gov.ph/index.php/programs-and-projects/innovations/256-project-star.

Gonzales, E. (2019, December 29). *YEAR-END Report: DepEd 2019: The Quest for Quality Education Continues.* Manila Bulletin. https://mb.com.ph/2019/12/29/year-end-report-deped-in-2019-the-quest-for-quality-education-continues/.

Holmlund, T., Lesseig, K., & Slavit, D. (2018). Making sense of "STEM education" in K-12 contexts. *International Journal of STEM Education, 5*(32). doi: https://dol.org/10.1186/s40584-018-0127-2

John, A., & Estonanto, J. (2017). Acceptability and difficulty of the stem track implementation in senior high school. *Asia Pacific Journal of Multidisciplinary Research, 5*(2), 43–50.

Karalis, T. (2016, June). Cascade approach to training: Theoretical issues and practical applications in non-formal education. *Journal of Education and Social Policy, 3*(2), 104–108. http://jespnet.com/journals/Vol_3_No_2_June_2016/12.pdf.

Lamberg, T., & Trzynadlowski, N. (2015). How STEM academy teachers conceptualize and implement STEM education. *Journal of Research in STEM Education, 1*(1), 45–58.

Lee, M. H., Chai, C. S., & Hong, H. Y (2019). STEM Education in Asia Pacific: Challenges and Development. *Asia-Pacific Education Researcher, 28*, 1–4. https://doi.org/10.1007/s40299-018-0424-z

Medenilla, S. (2019, September 23). *PHL Lags behind in Budget for Education in ASEAN.* Business Mirror. https://businessmirror.com.ph/2019/09/23/phl-lags-behind-in-budget-for-education-in-asean/.

Morales, M., Anito, J., Avilla, R. Abulon, E., & Palisoc, C. (2019). Proficiency indicators for Philippine STEAM (Science, Technology, Engineers, Agri/Fisheries, Mathematics) Educators. *Philippine Journal of Science, 148*(2), 263–275.

Nivera, G. (2018). *Science, Technology, Engineering and Mathematics (STEM) in the Philippines: An Education Pipeline Study. A National Research Project Funded by the USAID and the Philippine Business for Education.* Manila: PBED.

Nivera, G., Atweh, B., Golla, E., Butron, B., Avilla, R., & De Mesa, D. (2017). *Educators' Views on the Content and Pedagogy of Science and Mathematics Teacher Education Program for Junior Secondary Teachers: A Case Study of the Philippines. A University-funded Research Report Presented at the PNU Graduate Research Forum.* Manila: Philippine Normal University.

Novio, B. (2019, September 26). *Teachers Now Joining Diaspora of Filipinos Seeking Greener Pasture.* Inquirer.Net. https://globalnation.inquirer.net/180294/teachers-now-joining-diaspora-of-filipinos-seeking-greener-pasture#ixzz6uosfSgGT.

OSWG. (2020, October 27). *Gatchalian Seeks to Increase Centers of Excellence for Teacher Education.* Philippine Information Agency. https://pia.gov.ph/press-releases/releases/1057106.

Padolina, W. G. (2014, August 6). *Higher Education Science and Technology and Economic Competitiveness* [Powerpoint Slides]. CHED-HERRC National Research Conference, Davo City, Philippines. https://ched.gov.ph/wp-content/uploads/2017/11/Higher-Education-science-and-Technology-and-Economic-Competitiveness-%E2%80%93-Dr.-Padolina.pdf.

Penerio, F. T., & Toshihiko, M. (2020). Science teaching and learning in Japan and the Philippines: A comparative study. *Universal Journal of Educational Research, 8*(4), 1237–1245. doi: 10.13189/ujer.2020.080414.

Philippine Statistics Authority. (2020, June 4). *Total Number of OFWs Estimated at 2.2 M.* https://psa.gov.ph/statistics/survey/labor-and-employment/survey-overseas-filipinos.

(PISA) Programme for International Student Assessment. (2018). *PISA 2018 Results.* Organisation for Economic Co-operation and Development. https://www.oecd.org/pisa/publications./pisa-2018-results.htm.

Research Center for Teacher Quality (RCTQ) (2021, March). *RCTQ Assists PNU on Customized PD Programs for DepEd.* https://www.rctq.ph/?p=2454.

Reyes, R. (2019, January 14). *PBEd Calls for Overhaul of Teacher Education Programs.* Business Mirror. https://businessmirror.com.ph/2019/01/14/pbed-calls-for-overhaul-of-teacher-education-program/.

Salary Standardization Law of 2019. (2019) (Phl.). Retrieved from: https://www.official-gazette.gov.ph/downloads/2019/12dec/20200108-RA-11466-RRD.pdf.

Shulman, L. (1987). Knowledge and teaching: Foundations of the new reform. *Harvard Educational Review, 57*(1), 1–22.

Southwest Regional STEM Network. (2009). *Southwest Pennsylvania STEM Network Long Term Range Plan (2009–2018) Plan Summary.*

Tatto, M. T., Schwille, J., Senk, S. L., Ingvarson, L., Rowley, G., Peck, R., Bankov, K., Rodriguez, M., & Reckase, M. (2012). *Policy, Practice, and Readiness to Teach Primary and Secondary Mathematics in 17 Countries. Findings from the IEA Teacher Education and Development Study in Mathematics (TEDS-M).* Amsterdam, The Netherlands: International Association for the Evaluation of Student Achievement.

Teng, P. (2019, October 21). *STEM Education in a Changing Employment Landscape.* SciDevNet. https://www.scidev.net/asia-pacific/education/opinions/stem-education-in-changing-employment-landscape/.

Teodoro, L. (2020, July 9). *Philippine Education in Crisis.* Business World. https://www.bworldonline.com/philippine-education-in-crisis/.

Trends in International Mathematics and Science Study (TIMSS). (2019). *Trends in Mathematics and Science Study 2019.* International Association for the Evaluation of Educational Achievement. https://www.iea.nl/studies/iea/timss/2019.

Torre, R. (2017, October 3). *Group Flags 'Dismal' Record in Licensure Exam for Teachers.* Business World. https://www.bworldonline.com/group-flags-dismal-record-licensure-exam-teachers/.

Tupas, F. & Noderama, R. (2020). Looking into In-Service Training for Teachers in the Philippines: Are They Gearing Towards Education 4.0? *Universal Journal of Educational Research,8*(10),4651–4660. https://www.hrpub.org/download/20200930/UJER34-19516966.pdf.

World Bank. (2016). *Assessing Basic Education Service Delivery in the Philippines: The Philippine Public Education Expenditure Delivery Study.* World Bank Group and Australia Aid.

World Economic Forum. (2016). *The Future of Jobs: Employment, Skills and Workforce Strategy for the Fourth Industrial Revolution.* National Centre for Vocational Educational Research https://www.voced.edu.au/content/ngv:71706.

World Economic Forum. (2020). *The Future of Jobs 2020.* https://www.weforum.org/reports/the-future-of-jobs-report-2020.

5 STEM teacher education in Thailand

Chatree Faikhamta and Kornkanok Lertdechapat

The integration of Science, Technology, Engineering, and Mathematics (STEM) education has been prioritised as a major driving force that can help to move nations forward (Institute for Promotion of Science and Mathematics Teaching [IPST], 2014). STEM education can provide young people with the skills necessary to work in the 21st century (Bybee, 2011; NGSS Lead States, 2013). It can enhance students' higher-level thinking, creative thinking, problem-solving, and critical thinking skills, and STEM can also assist in the development of the communication skills and technological proficiency that can serve as tools of inquiry (Garibay, 2015; Gonzalez & Kuenzi, 2012). Hence, it is viewed as the lens of curriculum or pedagogy to promote mentioned aspects of students learning (Bybee, 2013; Saito et al., 2016).

Similar to other countries, Thailand is facing problems related to instruction in mathematics, science, and technology. The main concerns are as follows: (1) The declining number of students in the fields of science, mathematics, and technology at the basic, vocational, and higher education levels lead to the lack of a higher-level workforce in the future (Chalaemwong, 2014); (2) The results of the national and international evaluation indicate that the quality of science, mathematics, and technology education in schools is low (IPST, 2014); and (3) Now and in the future, Thailand requires a workforce with knowledge and skills in science and technology to be able to participate in highly competitive markets, such as modern agriculture, high technology, telecommunication, energy and environmental management, with information technology, high-technology machines, and logistics management. If nothing is done, Thailand will lose its potential competitiveness in the near future. Countries throughout the world are finding their own solutions to educational reform. Given the present situation in Thailand, all stakeholders of the education systems should find ways to improve the country's education. One possible solution throughout the world is to focus on enhancing STEM education, which allows teachers and students to augment the current curricula with current information and real-world events to make learning more dynamic, engaging, and valuable (Bryan et al., 2016).

The concept of STEM education in Thailand is similar to that in other countries (IPST, 2014). Its purpose is to help students understand and apply STEM competencies in their daily lives. The integration of all subjects and learning,

DOI: 10.4324/9781003099888-5

both in the classroom and in real life, is currently the primary trend in education to make learning more meaningful for students. Consequently, students can better realise the value of their studies and apply their knowledge in everyday life, both of which can lead to broader job opportunities in the future, add more value, and in turn, build up national economies. The Institute for the Promotion of Teaching Science and Technology (IPST) has called for increased studies in STEM fields across all levels of education. Teaching STEM in primary and secondary education can help students become interested in STEM careers and build a nation's STEM-educated workforce to meet business and industry demands in a complex and technology-driven economy. A new emphasis on STEM education that takes an integrated and multidisciplinary approach to education has gained prominence. This new approach caters to developing high-level thinking skills as they are applied in the real world by scientists, engineers, and other professionals to recognise, evaluate, and solve complex problems and discover new knowledge (IPST, 2013, 2014). Teachers can apply many approaches in their classrooms to develop 21st-century skills and prepare students to live in a rapidly changing world (Fortus et al., 2004). Doing so requires teachers to expose students to new knowledge and skills and inspire them to think and be creative (IPST, 2013).

Recently, Thailand has begun encouraging teachers to develop STEM-appropriate teaching practices via the Institute for the IPST's National Centre for STEM education agencies. The IPST (2013) proposes that STEM is "an approach that integrates science, engineering, technology, and mathematics, with a focus on solving real-life problems, including the development of new processes or products that benefit human life and work". According to IPST policy, science curriculum standards have been revised to comprise four strands of education (biological science, physical science, design and technology, and earth, astronomy, and space science), covering the 12 years of basic education.

The integration of STEM disciplines can involve at least two disciplines. For example, the science discipline should place the most emphasis on the engineering design process (EDP), which is the driver of STEM activities, while the other STEM disciplines should support students in completing the given challenges in STEM situations. These STEM disciplines and their relevant processes are the significant modes for enhancing students' skills in the real-life context. STEM-related practice can enhance students' learning and innovation skills. STEM education not only focuses on disciplines but also on the EDP, which is the mechanism of successful implementation of STEM education in the classroom.

The IPST (2015) introduced six indicators of STEM standards to help students apply the integration of STEM knowledge to solve real-life problems, including the following: (1) Identify problem; (2) Collect data and concepts relevant to the problem; (3) Design the solutions which are relevant to science, technology, engineering, and mathematics; (4) Plan and implement the solution; (5) Test, evaluate, and revise the solution; and (6) Communicate the solution and result from the test. The components of EDP are the key mechanisms of enhancing students' learning and innovation skills. The four core characteristics of integrated STEM instruction are similar to those proposed by Bryan et al. (2016), including the following: (1) The content and practices of one or more anchor science and mathematics disciplines define some of the learning goals; (2) The integrator is the practices of engineering and engineering design which provide real-world, problem-solving context for learning and meaningfully bring in other disciplines; (3) The engineering designs or engineering practices related to the relevant technologies require the development of scientific and mathematical concepts through design justifications, which involves recommending the design to the client and making decisions from the data; (4) The development of 21st-century skills is emphasised.

Thai teachers' understanding and implementation of STEM activities

STEM teachers across the world, including the teachers in Thailand, are challenged by the preparation of educational activities for STEM (Shernoff et al., 2017). The research on Thai teachers' understanding and implementation of STEM activities has increased. In a study by Faikhamta (2020), 428 pre-service science teachers who enrolled in science teacher education programmes have a naive understanding of the nature of STEM. The majority of pre-service teachers viewed science, technology, and mathematics as products and engineering as a process. They viewed science as relating to the things around them and knowledge that explains natural phenomena. They saw science as a search for the truth, rather than a means to construct explanations of natural phenomena. They seemed to believe that science is the application of technology and viewed technology as only consisting of artefacts, such as tools. Engineering was viewed as the design and construction of things, such as houses. This might indicate that teachers believe engineering must create a product, or a thing, rather than develop a process. They also thought that mathematics involves numerical calculations and collections of facts and rules. Moreover, pre-service science teachers' views on STEM integration were consistent with the view that "science and math are connected by technology and/or engineering" – one of the nine perspectives proposed by Bybee (2013).

Since teachers' perceptions directly affect their judgement of learning and teaching interactions in their classrooms and they, in turn, influence their classroom behaviour (Clark & Peterson, 1986), some researchers have investigated teacher's attitudes, perceptions, and beliefs about the integration of STEM (Sanders, 2009; Stohlmann et al., 2012; Vasquez et al., 2013). Teacher's perception is one of the PCK components that may function as a "filter and amplifier" through which teachers may screen their classroom experiences and interpret their subsequent classroom practices (Berry et al., 2008). Many studies have indicated that teachers still struggle to answer the question, "What is STEM education?", "How are STEM subjects integrated?" (Srikoom et al., 2017; Stohlmann et al., 2012; Vichaidit & Faikhamta, 2017).

Srikoom et al. (2017) illustrated a lack of consensus among teachers as to how STEM education is defined from an integrative perspective. No common definition of STEM has been agreed upon among in-service teachers. In-service teachers expressed interest in STEM, but they have concerns about teaching the various STEM disciplines, most notably engineering. Most of the teachers described STEM education as an integrated discipline teaching approach to solve real-life problems. They often defined STEM education by sharing the same phrases: "solving problems" and "learning by doing". Therefore, the types of activity that harmonise with Thai teachers' perceptions of STEM education are probably problem-based, project-based, and inquiry-based approaches. For the authentic problem-based preference, teachers view STEM as an approach that involved solving real-life problems with integrated knowledge. That is, real-life problems need to be infused in STEM lesson plans to connect lessons to everyday experiences and meet STEM education learning outcomes.

As aforementioned, science teachers have different conceptions of STEM disciplines (Vichaidit & Faikhamta, 2017). Many do not implement engineering design processes, and few integrate all four disciplines into the classroom (Srikoom et al., 2018). It can be argued that teachers lack the knowledge and experience needed to teach STEM and that most of them have limited understanding of how to include the appropriate perspectives of STEM integration in their teaching. Therefore, one of the biggest educational challenges in the implementation of K-12 STEM education is the lack of guidelines or models for teachers to follow that show how to use an integrated STEM approach in the classroom situation.

PCK-based STEM teacher education programme

In this section, we introduce pedagogical content knowledge (PCK) originally proposed by Shulman (1987) as a framework for developing teachers' competency in teaching STEM. PCK plays an important role in learning how to teach. It provides educators and researchers with an understanding of the process of instructors teaching a particular subject and learning how to teach science (Kind, 2009), through STEM, in this case. In the early years of the 21st century, PCK has become an important knowledge base for teachers and for revising teacher education programmes (Abell, 2008). Some researchers have used PCK as a framework in teacher education programmes to develop science teachers' teaching practice. The research indicates that increased PCK leads to greater teacher understanding of student learning difficulties and misconceptions, the enhancement of student learning, and the use of appropriate content-specific teaching strategies and multiple modes of representation (Berry et al., 2008; Geddis et al., 1993). Building on Shulman (1987)'s notion of PCK, many studies have defined PCK as consisting of instructional strategies that incorporate representations of subject matter and a good understanding of student learning difficulties and modes of thought across subjects (Gess-Newsome, 1999, 2015; Magnusson et al., 1999; Park & Oliver, 2008). Content knowledge (subject matter) alone is not sufficient to be an effective teacher, and thus, what teachers need is a deeper knowledge of PCK. Therefore, to help teachers teach STEM education in classrooms, Thailand must develop teachers' PCK to develop their capacities to teach the STEM approach.

Unfortunately, the development of PCK for teaching STEM represents a massive educational gap that Thailand needs to fill. Most research on teacher PCK has focused on how to educate teachers to transform their knowledge of content into appropriate methods of teaching (e.g., Appleton, 2003). Few authors have investigated the PCK of science teachers in teaching STEM in Thailand by considering the international standards or frameworks of STEM Education as the lens of their investigations.

Integrated STEM education lends itself to best practices for teaching, including the use of multiple representations, teacher as a facilitator, embedded formative assessment, cooperative learning, and problem-solving based learning (Stohlmann et al., 2012). From the definitions and elements of PCK first described and proposed by Schulman (1987), PCK in teaching STEM subjects (PCK-STEM) which was proposed by Lertdechapat (2020)'s work is developed as shown in Figure 5.1. PCK for STEM was conceptualised from the dimension of PCK and STEM education. In terms of PCK, teachers' knowledge and practices of particular concepts or topics are both supported through the backgrounds of teachers and students. The evidence from student outcomes would reflect how teachers' PCK is formed and delivered to students. Another dimension, STEM education is viewed as a way to develop students' learning and innovation skills through the process of engineering design.

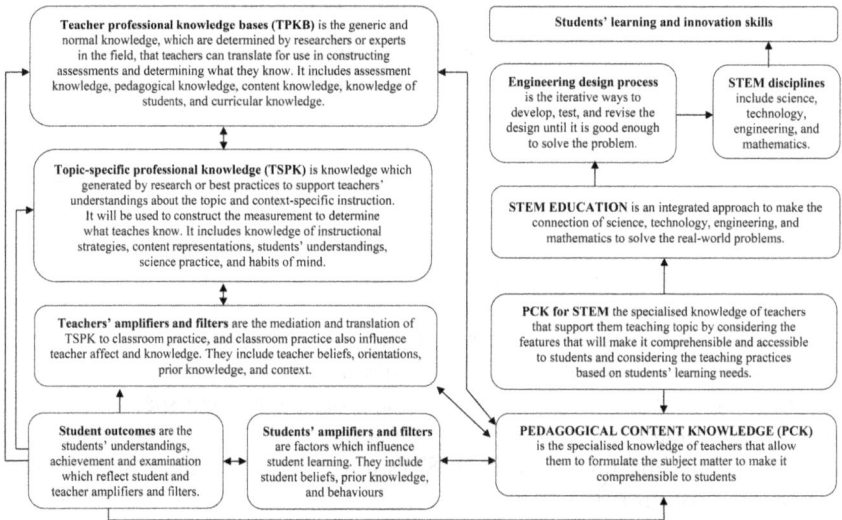

Figure 5.1 A theoretical framework of PCK for STEM.

Table 5.1 Pedagogical content knowledge for teaching STEM education (modified from Srikoom et al., 2018; Lertdechapat, 2020)

Components of PCK for STEM	Subcomponents of PCK for STEM
1 Orientations to teaching STEM	1.1 Teachers' knowledge and beliefs about the purposes and goals (orientations) for teaching and learning STEM
2 Knowledge of STEM curriculum	2.1 Curriculum materials used in meeting the goals and objectives of STEM concepts
	2.2 Educational standards related to STEM concepts
	2.3 Instructional goals in STEM concepts across disciplines
	2.4 Sequencing of topic within the broader unit of study
3 Knowledge of learners' understanding in STEM	3.1 Prerequisites: Knowledge and skills for learning a STEM lesson
	3.2 Common misconceptions about STEM concepts/students' learning
	3.3 Variations in approaches to learning the STEM concepts of particular topics
	3.4 Sources of students' difficulty and common errors about understanding STEM
4 Knowledge of instructional strategies for teaching STEM	4.1 Subject-specific strategies which relate STEM disciplines
	4.2 Topic-specific strategies: STEM activities
	4.3 Topic-specific strategies: Representations
	4.4 Strategies for adapting instructions for diverse students
5 Knowledge of assessment of STEM learning	5.1 What to assess
	5.2 Subject-specific assessment strategies
	5.3 Topic-specific assessment strategies
	5.4 Purpose of assessment

Since STEM education is considered as an instructional approach in which STEM disciplines are integrated (Bybee, 2013), science teachers who can teach STEM need specific knowledge to teach particular topics which can be incorporated into four STEM disciplines. Teachers' specialised knowledge and practices specific to STEM education have been explored in five dimensions including (1) Orientations to teaching STEM, (2) Knowledge of STEM curriculum, (3) Knowledge of students' understanding of STEM, (4) Knowledge of instructional strategies and representations for teaching STEM, and (5) Knowledge of assessment of STEM learning (modified from Srikoom et al., 2018; Lertdechapat, 2020) (See Table 5.1).

These PCK components for STEM reflect teachers' ability in STEM teaching practices that are the factors for implementing STEM education activities in the classroom, including how they conceptualise STEM, how they understand STEM curriculum and curriculum materials, what difficulties related to STEM students face, and student learning is assessed through STEM instruction. Thai

STEM teachers need to know, understand, and apply all five aspects of knowledge to design, implement, and effectively revise STEM activities in classrooms.

Because of its value for enhancing teachers' competency and students' learning, PCK for STEM can be used as a guide for designing science pre-service and in-service teacher education. We have elaborated the ideas of PCK for STEM and developed research studies (e.g., Srikoom et al., 2018) by providing supportive teacher education environments that are organised to support the personal and professional needs of teachers and value the input of teachers. Subsequently, teachers can gain a greater sense of ownership and be more inclined to accept and sustain STEM programme efforts. We also found that providing an example of an engineering design challenge for teachers to use as a reference model and demonstrating how to do engineering design and problem solving in and outside the classroom can enhance teachers' PCK for STEM. We recommend that teacher education designers and teacher educators need to have well-planned and continuous coaching and follow-up steps regarding PCK for STEM. Examples of using PCK for STEM framework are shown below in in-service teacher professional development and student teaching in pre-service teacher education.

Developing in-service science teachers' PCK for STEM through a professional development programme

The current body of literature suggests key features of professional development (PD) that can help in-service teachers develop their teaching ability. These include hands-on activities for teachers, modelling for teachers to apply and use in their class, the provision of opportunities to apply, observations and assessments in the classroom, sessions for teachers to share their ideas during participation, content linked to context, and coaching and follow-up (Asghar et al., 2012; Capobianco, 2011; Cochran-Smith & Lytle, 1993; Dare et al., 2014; El-Deghaidy et al., 2017; Nadelson et al., 2013; National Research Council [NRC], 2011; Rockland et al., 2010; Zembal-Saul et al., 2000). By following the literature, we arrived at guiding principles, specific to the context of science teachers who are responsible for implementing STEM lessons, to provide the basis for developing the necessary research intervention. The guiding principles include:

1   An in-service teacher is a learner. An individual STEM teacher is an adult learner who continues to develop his or her ideas about STEM, ways to teach STEM, and classroom activities. In this study, in-service teachers were viewed as learners, with their beliefs and knowledge taken into account.

2   An in-service teacher is a critical inquirer. The science teacher is a critical inquirer who displays scepticism of social phenomena, especially in classroom learning. Science teachers observe their STEM classroom and student learning, exhibit curiosity, and ask questions. They may design their own investigations to get the answers. The critical inquirer draws conclusions based on verified evidence before they share their conclusions with others.

3 An in-service teacher is a reflective practitioner. According to the assumption that knowledge is embedded in a socially structured context by a community of practitioners, a science teacher is a practitioner who promotes the exercise of analysis, judgment, and action. Through a research-based PD programme, science teachers are provided opportunities to develop their reflectivity in a variety of tasks, such as doing classroom action research, reflecting on their STEM teaching, and writing reflective journals and logbooks.

4 An in-service teacher is a critical friend. Science teachers are members of the learning community. They can inspire the development of ways of working with others that enable the kinds of social interaction necessary for renegotiating and reconstructing what it means to be a teacher. Through a research-based PD programme, teachers are provided opportunities to act as a critical friend who supports and gives feedback to others.

In a STEM professional development programme, teachers are provided with a supportive environment that is organised to support their personal and professional needs, and values the input of the teachers (Srikoom et al., 2018). Out of this, the teachers gain a greater sense of ownership and become more inclined to accept and sustain the STEM programme. Providing an example of an engineering design challenge for teachers to use as a reference model and demonstrating how to do engineering design and problem solving in and outside of the classroom can enhance teachers' PCK for STEM. Al Salami et al. (2017) suggest that programme developers should also provide opportunities for teachers to enhance their teamwork skills. Nadelson et al. (2013) found that the provision of resources and engaging teachers in hands-on, inquiry-based activities could support the teachers in developing a deeper understanding of STEM. Zembal-Saul et al. (2000) integrated planning, teaching, and guided reflection into cycles of instruction in a methods course designed to develop science teachers' PCK. They argued that teachers should be provided with the chance to plan, teach, and reflect on their teaching. This has the potential to provide teachers with an opportunity to revise, reformulate, and refine their beliefs about teaching and learning.

In the study by Faikhamta et al. (2020), a PCK-based STEM professional development programme was designed based on various strategies with specific objectives. The objectives were to (1) Assist teachers with understanding the nature of STEM subjects and their integration, (2) Enhance teachers' ability to navigate their own STEM teaching approaches, (3) Support teachers in the integration of STEM into science content in their own classroom, and (4) Support teachers' positive attitudes towards STEM. To address the goal, action research was a key vehicle to help in-service teachers develop their teaching ability. Action research is "first-person research", in which the researcher is the researched, aiming to develop, improve, and understand his or her situation (Capobianco, 2011; Kemmis, 1991). Because action research is a socially constructed activity involving the discussion and dissemination of results to others, it offers teachers opportunities to construct or reconstruct their own knowledge and beliefs through collaboratively working with colleagues (Cochran-Smith & Lytle, 1993; McNiff, 1993). Through action research, in-service teachers have a chance to plan lessons, teach, and reflect on teaching, and these are key concepts in developing science teachers' PCK, both in terms of understanding and practice (Bertram & Loughran, 2012). Zembal-Saul et al. (2000) integrated planning, teaching, and guided reflection into cycles of instruction in a methods course designed to promote science teachers' PCK. They argued that each element should not be isolated and that cycles of instruction should be dynamic. As mentioned in the literature, one of the main concerns is how to integrate STEM into the content being taught in their classes. Since the teachers had a set of skills and topics from particular domains (physics, chemistry, and biology), classroom action research with support by programme developers would help them find their own ways of teaching.

The intervention consisted of several stages and involved different activities (See Table 5.2). First, the participants were exposed to argumentation about the nature of STEM subjects, integrated STEM, and teaching STEM through sharing experiences and engaging STEM activities conducted by a researcher. The participants were asked to discuss problems in teaching STEM in their classroom in groups of five or six. Then, participants were put into groups based on their common problems and interests, including students' creative thinking, critical thinking, EDP skills, collaborative problem skills, scientific conceptions, and scientific argumentation. After the participants were placed in different groups, they were required to develop a common research question.

The in-service teachers were trained about classroom action research. The sessions included hands-on activities and a short PowerPoint presentation highlighting the different aspects of classroom action research and conventional research. For instance, we introduced the participants to different classroom action research paradigms (McNiff, 1993). The framework helped the participants better understand the objectives, importance, and process of classroom action research. In addition, we engaged them to criticise, compare, and contrast articles from action research and conventional research. Following the training session, the groups were given four hours to develop a classroom research proposal that was consistent with their common problem, including data collection and analysis.

Then, the groups were given four hours to develop their intervention and a STEM lesson that was consistent with their research proposal. The participants were provided with sample STEM lessons that could be used to design their own STEM lessons. Each group was asked to present their STEM lesson to an entire class through a gallery walk. The purpose of this activity was to provide the scaffolding needed for the participants to develop their lessons. During the gallery walk, in-service teachers were in constant communication with their peers and were provided feedback when needed. After this planning section, everyone was asked to plan their own STEM lessons and implement them. The lesson implementations lasted for six weeks.

Table 5.2 Outline of PD activities

Sessions	Content	Activities
1	• Nature of science, technology, engineering, and mathematics • Goals of teaching and learning STEM	• Examine the nature of S/T/E/M activities • Perspectives in teaching STEM • Share ideas about problems in teaching practices in real settings
2	Experiences of teaching science	• Discuss and make conclusions about their roles as science teachers • Articulate problems and learn about their teaching STEM • Watch video recording of teaching STEM • Reflect on scenarios in the video
3	Classroom action research	• Draw on research questions based on their problems in teaching STEM • Propose classroom action research design • Share and discuss new paradigms of classroom action research • Share their action research plan
4	Lesson plans and new teaching strategies	• Share ideas about teaching in terms of the goals of teaching STEM, students' understanding and learning about specific topics, teaching strategies, and the assessment of students' learning • Design STEM lesson plans to be implemented in real context
5	Implementation of STEM lesson	• Try new STEM lesson plan
6	Reflecting on STEM implementation and sharing classroom action research results	• Bring all data from STEM implementation • Discuss data analysis and write research results • Present and discuss results of STEM lesson implementation and classroom action research

Both the course instructor and the peers provided feedback after each implementation through Google Classroom. Doing so helped the in-service science teachers develop their own teaching style and data collection approach. Each teacher was required to collect data on student learning using the research tools they developed and data on their own teaching practice using reflective journals. The purpose of this requirement was to help the participants identify weaknesses and strengths in students' learning and to help them develop their teaching skills in STEM.

Finally, in-service teachers came back to the university for a session to learn how to analyse the collected data and write a research report. The focus of this session was to help them write a narrative explaining their own learning, changes in student learning, and changes in their STEM teaching practice.

Then they shared their own research results through a gallery walk activity. The course instructor and the peers also provided feedback, which could help them revise their research reports.

After the programme, the teachers had a more positive attitude towards STEM education. They had a better understanding of how to balance the development of students' science concepts with the engineering process. They already had a well-developed and adequate understanding of STEM. Science teachers had a better understanding of the nature of the STEM disciplines and especially the nature of engineering, which was a key factor in helping them understand the STEM teaching approach (Nadelson et al., 2013; Roehrig et al., 2012). Prior to the intervention, most of the teachers had a limited understanding of engineering. They became aware that engineering is not only about making artefacts and designing things, but it is also the way engineers think and approach solving problems. We found that providing an example of an engineering design challenge for teachers to use as a reference model and demonstrating how to do engineering design and problem solving in and outside the classroom could enhance the teachers' ability to teach STEM.

Developing pre-service teachers' PCK for STEM through lesson study in the context of their student teaching

Preparing pre-service teachers for teaching STEM is one of the main efforts in science teacher education. Since science curriculum in basic education has become more STEM-oriented, teacher educators should design courses and activities to help pre-service teachers to become STEM teachers. Compared to in-service teachers, pre-service teachers would have more serious problems than in-service teachers. For example, they had less confidence in carrying out the projects in STEM lessons (Siew et al., 2015), and they had low levels of understanding of the nature of science, which is a part of STEM disciplines (Buaraphan, 2010). In this chapter, we present a case study of helping pre-service teachers practice STEM lessons in the context of student teaching.

We use lesson study (LS) as a potential strategy to develop pre-service science teachers' PCK for STEM. LS is one of the recommended strategies to enhance pre-service science teachers' ability to develop their PCK for STEM (Lertdechapat, 2020). During field experience, pre-service science teachers are provided a platform to practice and gain insightful teaching experience (Varma & Hanuscin, 2008). However, they still confronted problems in implementing the teaching science and pedagogical strategies (Moore, 2003; Varma & Hanuscin, 2008) and had a deficit of beliefs in teaching science (Bhattacharyya et al., 2009). Since the teachers in school have diverse levels of STEM teaching experience as well as different points of view from different disciplines, collaboration among those teachers is required for enhancing teacher candidates in PCK for STEM.

Although a call has been made for pre-service science teachers to be critical inquirers and to develop their metacognition in teaching and discussions on student learning (Faikhamta et al., 2018), the collaboration among cooperating teachers, university mentors, and pre-service science teachers in terms of reflection on student learning is not quite emphasised. This claim was explicitly referred to in the previous study in which student learning was not guaranteed because of collaboration between pre-service science teachers and cooperating teachers who were specialists in science instruction (Varma & Hanuscin, 2008). In addition, an emphasis on teachers' specialised knowledge is required (Faikhamta, 2013). The activities at each step of LS for enhancing teachers' PCK for STEM (Lertdechapat, 2020) are shown in Table 5.3. The main idea of Table 5.3 was modified from Lewis and Hurd (2011)'s study which conceptualized the LS as a means for enhance professional learning community. Although LS is expected to enhance teachers' PCK for STEM, the targeted PCK for STEM components specific to each phase could not be predetermined but the LS members should consider the relevant components when they follow the activities of each phase.

To determine the aspects of students' learning, which is the goal of LS, the problems students confront and how their learning success could be conducted in the authentic context are explored. Prior to implementing the LS, pre-service science teachers were asked to observe students' problems and success while engaging in the science lessons. Then, the pre-service science teachers shared their observations to the LS cluster including a pre-service science teacher, his or her cooperating teacher, and university mentor, brainstormed the possible sources, and selected the most significant problem or struggle which could be solved by the implementation of STEM lessons. The selected problem is referred to as the goal of LS which will be developed through the study process. To prepare the well-developed teaching strategies, the relevant information regarding the aspect of students' learning, the instructional strategies to implement the STEM lessons, and the strategies to assess students' learning were systematically developed by all the LS members. The powerful collaboration among the pre-service science teachers, the cooperating teachers, and the university mentors will generate the robust knowledge and means to conduct the LS in their context of science lessons (Cajkler & Wood, 2015).

Lertdechapat (2020) found that the most significant process of LS is the post-lesson discussions among the members after implementing the lessons. Learning from the knowledge of others can enhance pre-service science teachers' ability to develop their PCK for STEM. According to the LS enactments, the pre-service science teachers were encouraged to discuss students' learning with their cooperating teachers and university mentors. Thus, they could learn from the post-lesson discussions rather than gain experience through the accumulation of the LS engagement.

Most subcomponents of PCK for STEM could be discussed during post-lesson discussions, and this platform could help pre-service science teachers to develop their understanding and practice relevant to the specific PCK for STEM. Regarding their actual post-lesson discussions, ten categories, including STEM lesson goals, prototypes, the student context, content, teaching, visualisation,

Table 5.3 Activities at each step of LS for enhancing teachers' PCK for STEM (Lertdechapat, 2020)

Phases of lesson study	Steps of lesson study	Activities	
1 Revisit the students' learning	1.1 Exploring the struggles and success of students' learning when they engage in normal science lessons	1.1.1	The PSTs and CTs observe the struggles and success as well as the students' learning behaviour when they engage in normal science classes.
		1.1.2	Observations of the students' learning would be recorded and then presented to the meeting of all LS clusters.
		1.1.3	The LS clusters will discuss the possible sources of the mentioned students' struggles and success and finalise the students' learning which would be developed through the STEM lessons.
		1.1.4	Each cluster will study the framework of the selected students' learning and STEM instructional strategies.
2 Plan	2.1 Choosing a topic and formulating learning goals	2.1.1	Select the content which could be taught through three rounds of STEM instructional strategies within a semester.
	2.2 Designing the research lesson	2.2.1	Design the lesson to include planning the overall activities by responding to the content representation tools (CoRes).
		2.2.2	Design the lesson plans based on the completed CoRes.
	2.3 Designing how to collect data	2.3.1	Design the tools for collecting the particular students' learning which is studied in 1.1.4
3 Do and See	2.4 Making an appointment	2.4.1	Appoint the date, time, and room for observing the students' learning.
	3.1 Teaching and observing the research lesson	3.1.1	One teacher (either pre-service teacher or cooperating teacher) conducts the lesson which is designed in 2.2.2
		3.1.2	Other members of the cluster observe the students' learning using the decided framework of the students' learning and record field notes individually.
4 Revise	4.1 Reflecting how well students learn the researched lesson	4.1.1	Organise their own post-lesson discussions to reflect the students' learning based on their own observations.
	4.2 Preparing the second lesson	4.2.1	Revise the lesson plan based on post-lesson discussions in 4.1.1.
5 Reteach	5.1 Reteaching and observing the revised lesson	5.1.1	The revised lesson plan is conducted by another teacher (either pre-service teacher or cooperating teacher) while the others observe the classroom.
	5.2 Reflecting on the revised lesson	5.2.1	Reflect on the students' learning which results from the revised lesson and how the modification affects the students' learning.

student learning, logistics, and LS goals, were discussed during the post-lesson discussions. Among them, the two following subcomponents of PCK for STEM were much enhanced from the post-lesson discussions.

When the LS clusters discussed the goals of STEM lessons, prototypes, the student context, content, and teaching, the pre-service science teachers could explicitly develop their understanding and practice in regard to topic-specific strategies for STEM activities. Similar to the student difficulties and common errors subcomponents, the pre-service science teachers could develop their understanding and practice in this subcomponent when their LS cluster discussed the topics of prototypes, content, teaching, student learning, and logistics.

Although some categories occurred in both the mentioned subcomponents of PCK for STEM, the details of post-lesson discussions were different. In terms of the prototypes, the main ideas of the discussions reflected different purposes which link to different subcomponents of PCK for STEM. After the lesson, the cluster discussed how students could design the prototypes which were affected by the sequence of introducing the criteria and constraints. In the modified lesson, the students were asked to revisit the criteria and constraints before continuing to design and build their toys. This shows the teacher's understanding of the EDP, which was used as a teaching strategy for a lesson on the properties of materials.

Another example of the post-lesson discussions involved the possibilities of building prototypes. The pre-service science teacher found that some ideas and materials for building the toys might be too complicated for them to apply in the activity. This pre-service science teacher was concerned that the fourth graders could not use a drone and blades for making the toys fly. Furthermore, it might be difficult for students to understand how a drone works and how it could be applied in building the toys. This issue explicitly showed a concern about the students' difficulties and errors in designing and building the toys.

Last but not least, engaging in post-lesson discussions could not ensure the development of PCK for STEM unless the pre-service science teachers implemented what they discussed in their own lessons. This practice explicitly confirms that PCK for STEM involves not only the teachers' understanding but also the acts or practices of the PCK. Gess-Newsome (2015) described PCK as PCK & S, which refers to the act of teaching instead of the knowledge of teaching.

Concluding remarks

This chapter emphasises the idea that PCK for STEM is considered as the act of teaching STEM to achieve the goal of STEM education. To develop well-prepared pre-service and in-service STEM teachers, educators should understand the nature of STEM and be able to implement STEM lessons in specific contexts. Teachers should understand the common features of STEM disciplines. Kelley and Knowles (2016) suggested that a conceptual understanding of integrated STEM education must be a prerequisite for teaching. Pre-service teachers and in-service teachers should understand seven crosscutting concepts (NGSS Lead States, 2013) patterns; (1) Cause and effect, (2) Scale, (3) Proportion and quantity, (4) Systems and system models, (5) Energy and matter, (6) Structure and function, and (7) Stability and change. These crosscutting concepts reflect the commonality and integrative features of STEM disciplines; for example, when scientists observe the natural phenomena, they look for and present the patterns using the mathematical representations to illustrate the observed phenomena and increase the precision of their investigations. This example shows that scientific practices are incorporated with relevant STEM disciplines, not only the body of knowledge but also the nature of each discipline. Consequently, teacher education programmes should initially provide pre-service teachers with opportunities to develop their understanding of the nature of STEM and help them to deliver that understanding to their students in classrooms. In these courses, pre-service teachers should have the chance to analyse the nature of STEM in science curricula, consider students' conceptions of the nature of STEM, assess students' understandings of the nature of STEM, and learn how to make certain features of the nature of STEM explicit to students.

Teacher education programmes should emphasise teachers' understanding of STEM education and how to develop and implement STEM lessons in the classroom. We recommend that a PCK-based STEM professional development programme is an important intervention to guide teacher educators in developing their teachers' ability to teach STEM. Teacher educators can bring key features of the PCK-based STEM PD programme into their context. As learners construct and reconstruct their own understanding within a particular social context, the designers of this intervention should take the context into account. Importantly, to encourage Thai teachers to teach effectively, science teachers should be critical inquirers, reflective practitioners, and critical friends. At the policy level, implementing a PCK-based STEM PD programme is feasible and can benefit science teachers. The programmes were not designed for teachers to solely learn the STEM approach but also the PCK for STEM for specific content and contexts. Particularly in Thailand, there were some STEM workshops that were designed to model STEM lessons without bringing particular content and the students they teach into account. Teacher PD programmes should provide teachers with opportunities to learn in a context that reflects their PCK for STEM. They should have a chance to share and reflect on their prior knowledge about the nature of each STEM discipline, their integration, STEM learning goals, STEM teaching ideas, their learning approaches and struggles, and their own new STEM teaching strategies in the practical context. They should also be able to implement their lessons and reflect on their lessons.

To help them develop PCK for STEM in the teacher preparation programme, diverse points of view from experienced teachers and educators are needed. The courses specific to STEM education, such as the STEM methods course, should provide opportunities for the learners to gain knowledge from the people who are experienced in all aspects of STEM, and they should be encouraged to design and conduct STEM education by themselves. The suggestions from those experienced people would reflect their PCK for STEM in terms of understanding and enactment.

In the case of student teaching, the environment of LS should be settled in the school context as usual. A collaboration among teachers who may have diverse experiences with teaching STEM would be a significant means to promote in-service teachers' PCK for STEM during their normal schoolwork. Instead of attending a workshop outside their school, which may be not relevant to their school context, learning from each other within their unique context would be a practical approach to professional teacher development.

References

Abell, S. K. (2008). Twenty years later: Does pedagogical content knowledge remain a useful idea? *International Journal of Science Education*, *30*(10), 1405–1416.

Al Salami, M. K., Makela, C. J., & De Miranda, M. A. (2017). Assessing changes in teachers' attitudes toward interdisciplinary STEM teaching. *International Journal of Technology and Design Education*, *27*(1), 63–88.

Appleton, K. (2003). How do beginning primary school teachers cope with science? Toward an understanding of science teaching practice. *Research in Science Education*, *33*(1), 1–25.

Asghar, A., Ellington, R., Rice, E., Johnson, F., & Prime, G. M. (2012). Supporting STEM education in secondary science contexts. *Interdisciplinary Journal of Problem-Based Learning*, *6*(2), 85–125.

Berry, A., Loughran, J., & van Driel, J. H. (2008). Revisiting the roots of pedagogical content knowledge. *International Journal of Science Education*, *30*(10), 1271–1279.

Bertram, A., & Loughran, J. (2012). Science teachers' views on CoRes and PaP-eRs as a framework for articulating and developing pedagogical content knowledge. *Research in Science Education*, *42*(6), 1027–1047.

Bhattacharyya, S., Volk, T., & Lumpe, A. (2009). The influence of an extensive inquiry-based field experience on pre-service elementary student teachers' science teaching beliefs. *Journal of Science Teacher Education*, *20*(3), 199–218.

Bryan, L. A., Moore, T. J., Johnson, C. C., & Roehrig, G. H. (2016). *Integrated STEM education. STEM Road Map*. New York: Routledge.

Buaraphan, K. (2010). Pre-service and in-service science teachers' conceptions of the nature of science. *Science Educator*, *19*(2), 35–47.

Bybee, R. W. (2011). Scientific and engineering practices in K-12 classrooms. *Science Teacher*, *78*(9), 34–40.

Bybee, R. W. (2013). *The Case for STEM Education: Challenges and Opportunities*. Virginia: NSTA.

Cajkler, W., & Wood, P. (2015). Lesson study in initial teacher education. In P. Dudley. (Ed.) *Lesson Study: Professional Learning for Our Time*. New York: Routledge.

Capobianco, B. M. (2011). Exploring a science teacher's uncertainty with integrating engineering design: An action research study. *Journal of Science Teacher Education*, *22*(7), 645–660.

Chalaemwong, Y. (2014). *The Future of Workforce Market during COVID-19 by Thailand Development Research Institute* [Online]. Retrieved from: https://tdri.or.th/2020/03/labor-market-covid19/, March 30, 2020.

Clark, C., & Peterson, P. (1986). Teachers' thought processes. In M. Wittrock (Ed.), *Handbook of Research on Teaching* (pp. 255–298). New York: Macmillan.

Cochran-Smith, M., & Lytle, S. L. (1993). *Inside/Outside: Teacher Research and Knowledge.* New York: Teachers College Press.

Dare, E. A., Ellis, J. A., & Roehrig, G. H. (2014). Driven by beliefs: Understanding challenges physical science teachers face when integrating engineering and physics. *Journal of Pre-College Engineering Education Research (J-PEER), 4*(2), 47–61.

El-Deghaidy, H., Mansour, N., Alzaghibi, M., & Alhammad, K. (2017). Context of STEM integration in schools: Views from in-service science teachers. *EURASIA Journal of Mathematics, Science & Technology Education, 13*(6), 2459–2484.

Faikhamta, C. (2013). The development of in-service science teachers' understandings of and orientations to teaching the nature of science within a PCK-based NOS course. *Research in Science Education, 43*(2), 847–869.

Faikhamta, C., Ketsing, J., Tanak, A., & Chamrat, S. (2018). Science teacher education in Thailand: A challenging journey. *Asia-Pacific Science Education, 4*(1), 1–18.

Faikhamta, C. (2020). Pre-service science teachers' views of the nature of STEM. *Science Education International, 31*(4), 356–366.

Faikhamta, C., Lertdechapat, K., & Prasoblarb, T. (2020). The Impact of a PCK-based professional development program on science teachers' ability to teaching STEM. *Journal of Science and Mathematics Education in Southeast Asia, 43.* Retrieved from: http://myjms.mohe.gov.my/index.php/jsmesea/article/view/10145.

Fortus, D., Dershimer, R. C., Krajcik, J., Marx, R. W., & Mamlok-Naaman, R. (2004). Design-based science and student learning. *Journal of Research in Science Teaching, 41*(10), 1081–1110.

Garibay, J. C. (2015). STEM students' social agency and views on working for social change: Are STEM disciplines developing socially and civically responsible students? *Journal of Research in Science Teaching, 52*(5), 610–632.

Geddis, A. N., Onslow, B., Beynon, C., & Oesch, J. (1993). Transformation content knowledge: Learning to teaching isotopes. *Science Education, 77*(6), 575–591.

Gess-Newsome, J. (1999). Pedagogical content knowledge: An introduction and orientation. In *Examining Pedagogical Content Knowledge* (pp. 3–17). Dordrecht: Springer.

Gess-Newsome, J. (2015). A model of teacher professional knowledge and skill including PCK: Results of the thinking from the PCK summit. In A. Berry, P. Friedrichsen, & J. Loughran (Eds.), *Re-examining Pedagogical Content Knowledge in Science Education* (pp. 28–42). New York, NY: Routledge.

Gonzalez, H. B., & Kuenzi, J. J. (2012). *Science, Technology, Engineering, and Mathematics (STEM) Education: A Primer.* Washington, DC: Congressional Research Service, Library of Congress.

Kelley, T. R., & Knowles, J. G. (2016). A conceptual framework for integrated STEM education. *International Journal of STEM Education, 3*, 1–11.

Kemmis, S. (1991). Critical education research. *Canadian Journal for the Study of Adult Education, 5*, 94–119.

Kind, V. (2009). Pedagogical content knowledge in science education: Perspectives and potential for progress. *Studies in Science Education, 45*(2), 169–204.

Lertdechapat, K. (2020). *The Development of Teachers' Pedagogical Content Knowledge for STEM Teaching through Lesson Study to Enhance Students' 21st Century Learning and Innovation Skills* [Doctoral dissertation, Kasetsart University].

Lewis, C., & J. Hurd. (2011). *Lesson Study Step by Step: How Teacher Learning Communities Improve Instruction*. Portsmouth, NH: Heinemann.

Magnusson, S., Krajcik, J., & Borko, H. (1999). Nature, sources, and development of pedagogical content knowledge for science teaching. In *Examining Pedagogical Content Knowledge* (pp. 95–132). Dordrecht: Springer.

McNiff, J. (1993). *Teaching as Learning: An Action Research Approach*. London: Routledge.

Moore, R. (2003). Reexamining the field experiences of preservice teachers. *Journal of Teacher Education*, 54(1), 31–42.

Nadelson, L. S., Callahan, J., Pyke, P., Hay, A., Dance, M., & Pfiester, J. (2013). Teacher STEM perception and preparation: Inquiry-based STEM professional development for elementary teachers. *The Journal of Educational Research*, 106(2), 157–168.

National Research Council. (2011). A framework for K-12 science education: Practices, crosscutting concepts, and core ideas. *Committee on a Conceptual Framework of New K-12 Science Education Standards. Board on Science Education. Division of Behavioral and Social Sciences and Education*. Washington, DC: The National Academies Press.

NGSS Lead States. (2013). *Next Generation Science Standards: For States, by States*. Washington, DC: The National Academy Press.

Park, S., & Oliver, J. S. (2008). Revisiting the conceptualisation of pedagogical content knowledge (PCK): PCK as a conceptual tool to understand teachers as professionals. *Research in Science Education*, 38(3), 261–284.

Rockland, R., Bloom, D. S., Carpinelli, J., Burr-Alexander, L., Hirsch, L. S., & Kimmel, H. (2010). Advancing the "E" in K-12 STEM education. *Journal of Technology Studies*, 36(1), 53–64.

Roehrig, G. H., Moore, T. J., Wang, H. H., & Park, M. S. (2012). Is adding the E enough? Investigating the impact of K-12 engineering standards on the implementation of STEM integration. *School Science and Mathematics*, 112(1), 31–44.

Saito, T., Anwari, I., Mutakinati, L., & Kumano, Y. (2016). A look at relationships (Part I): Supporting theories of STEM integrated learning environment in a classroom – A historical approach. *K-12 STEM Education*, 2(2), 51–61.

Sanders, M. (2009). STEM, STEM education, STEM mania. *The Technology Teacher*, 68(4), 20–26.

Shernoff, D. J., Sinha, S., Bressler, D. M., & Ginsburg, L. (2017). Assessing teacher education and professional development needs for the implementation of integrated approaches to STEM education. *International Journal of STEM Education*, 4(1), 1–16.

Shulman, L. S. (1987). Knowledge and teaching: Foundations of the new reform. *Harvard Educational Review*, 57(1), 1–22.

Siew, N. M., Amir, N., & Chong, C. L. (2015). The perceptions of pre-service and in-service teachers regarding a project-based STEM approach to teaching science. *SpringerPlus*, 4(1), 1–20.

Srikoom, W., Faikhamta, C., & Hanuscin D. L. (2018). Dimensions of effective STEM integrated teaching practice. *K-12 STEM Education*, 4(2), 313–330.

Srikoom, W., Hanuscin, D. L., & Faikhamta, C. (2017). Perceptions of in-service teachers toward teaching STEM in Thailand. *Asia-Pacific Forum on Science Learning and Teaching*, 18(2), 1–23.

Stohlmann, M., Moore, T. J., & Roehrig, G. H. (2012). Considerations for teaching integrated STEM education. *Journal of Pre-College Engineering Education Research (J-PEER)*, *2*(1), 28–34.

The Institute for the Promotion of Teaching Science and Technology [IPST]. (2013). *STEM Instructional Strategies.* Retrieved from: http://physics.ipst.ac.th/?page_id=2481, March 12, 2020.

The Institute for the Promotion of Teaching Science and Technology [IPST]. (2014). *What is STEM Education?* Retrieved from: http://www.stemedthailand.org, September 12, 2019.

The Institute for the Promotion of Teaching Science and Technology [IPST]. (2015). *Standards of STEM Education.* Bangkok: Success Publication.

Varma, T., & Hanuscin, D. L. (2008). Pre-service elementary teachers' field experiences in classrooms led by science specialists. *Journal of Science Teacher Education*, *19*(6), 593–614.

Vasquez, J., Schneider, C., & Comer, M. (2013). *STEM Lesson Essentials, Grades 3–8: Integrating Science, Technology, Engineering, and Mathematics.* Portsmouth, NH: Heinemann.

Vichaidit, C., & Faikhamta, C. (2017). Exploring orientations toward STEM education of preservice science teachers. *Rajaphat Mahasarakham University Journal*, *11*(3), 165–174.

Zembal-Saul, C., Blumenfeld, P., & Krajcik, J. (2000). Influence of guided cycles of planning, teaching, and reflection on prospective elementary teachers' science content representations. *Journal of Research in Science Teaching: The Official Journal of the National Association for Research in Science Teaching*, *37*(4), 318–339.

Section 3

STEM curriculum

Context, challenges, and promises

Aik-Ling Tan

Introduction

This commentary raises some questions about science, technology, engineering, and mathematics (STEM) curriculum as a primer for further discussions for STEM education scholars, curriculum developers, policy makers, and teachers. Based on the ideas shared by Edwehna Elinore S. Paderna and Sheryl Lyn C. Monterola (Chapter 7), Si Qi Toh, Tang Wee Teo, and Yann Shiou Ong (Chapter 8), Ida Ah Chee Mok and Zhipeng Ren (Chapter 6) and Tenzin Sherab, Kinley Kinley, and Wangchuk Tempa (Chapter 9) in this volume, I compared the key purposes of integrated STEM education in Singapore, the Philippines, Hong Kong, and Bhutan to anchor my discussions on the purposes of curriculum design, context, challenges, and promises of integrated STEM education in the four economies.

The relevance and importance of STEM education to develop 21st-century competencies in students have been discussed greatly. Education systems are relying on successful STEM education to produce the next generation of workforce that is innovative, technologically literate, is flexible, adaptable, has initiative, is self-directed, and has strong social and cultural awareness. Consequently, it is important that the design of STEM curriculum takes into consideration 21st-century competencies and the needs of the workforce. Beyond 21st-century competencies, Honey et al. (2014) highlighted in their review the tensions that exist between teaching and learning of STEM as mono-discipline as compared with conceptualising STEM as an integrated discipline. The deep-rooted tradition and practice of presenting each discipline as a unique entity with its own knowledge base, specialised practices, and habits of mind, make integration across disciplines a challenge. Purists guard, argue against and warn that integration will lead to dilution of learning of disciplinary knowledge. Advocates for integrated STEM argue for learning in a more connected manner across disciplines with an emphasis on application in solving real-world problems.

DOI: 10.4324/9781003099888-III

The dichotomous views of discipline boundaries are compounded by issues such as STEM teacher identity, school structures, and STEM teacher education. Teacher education in many countries is still organised largely around expert knowledge within mono-disciplines. A teacher is developed to be a science teacher, a mathematics teacher, or a design teacher. Very rarely are teacher education programmes designed such that an individual is trained to teach all the four disciplines of STEM. Consequently, teachers engaged in teaching integrated STEM may not feel adequately prepared to handle the knowledge and skills of the other disciplines. While it is argued that integrated STEM needs to be implemented by a group of teachers rather than a single teacher, the current structure of schools is not ideal for team teaching. School schedules are still largely organised around single disciplines. There is usually no specific subject or time set aside especially for STEM lessons. The peculiar features of integrated STEM also led to scholars arguing that teachers engaged in STEM teaching and learning should be identified as STEM teachers or teachers of STEM.

Against the backdrop of ongoing scholarly discourse and tensions pertaining to integrated STEM education, efforts have been made across different education systems to try to actualise integrated STEM experiences for learners. Through designing bespoke STEM programmes, Paderna and Monterola, Mok and Ren, Sherab et al., and Toh et al., describe integrated STEM curricula in four different yet similar social contexts in the Philippines, Hong Kong, Bhutan, and Singapore. Paderna and Monterola (Chapter 6) described a partnership between Unilab Foundation and the University of the Philippines College of Education to set up the Centre for Integrated STEM Education, Inc. (CISTEM). The CISTEM, established in 2019 by Unilab Foundation's STEM+PH in partnership with the University of the Philippines College of Education, aimed to strengthen integrated STEM education through capacity building for teachers and educational institutions, curricular innovations, maximisation of network linkages, and learner empowerment. Targeting teacher professional development, the programme aimed to help teachers learn about design thinking in a STEAM programme. Toh et al. (Chapter 7) presented various STEM-based educational programmes (for example applied learning programme) in Singapore and used Rasch to examine students' views, attitudes, identity, self-concept, and career decisions after experience with integrated STEM programmes. Mok and Ren (Chapter 8) provided insights into school-base STEM curriculum that considers both contextual and personal aspects of learning. In the absence of a formal integrated STEM curriculum and initiative nationally, school-based initiatives introduced by enthusiastic teachers could form the impetus for STEM curriculum reforms. Finally, Sherab et al. (Chapter 9) presented insights into Bhutan's effort to shift from didactic mono-disciplinary STEM learning to one that is student-centric and incorporates scientific inquiry. They presented a case of how Grade 12 students' conceptual understanding of gene expression improved when inquiry forms of learning are introduced.

Purposes of STEM programmes

The authors from the Philippines, Singapore, Hong Kong, and Bhutan all referred to preparing a skilled and educated workforce as the context for STEM curriculum. Toh, Teo, and Ong quoted the Prime Minister of Singapore who positioned STEM capabilities as fundamental to "upgrading of Singapore's economy" (Toh et al.). Singaporean students are well-positioned to excel in STEM since Singaporean students topped the OECD's PISA global competence test in 2018 (Schleicher, 2019). As such, investing in STEM capabilities for the next lap of economic growth for the country is leveraging on an area of strength. Similarly, Paderna and Monterola mentioned "nation-building" and "economic prosperity" as reasons for the Philippines to embark on STEM education. They also stated that as a country, they possess several pre-requisites for successful STEM education – Filipinos being recognised internationally for entrepreneurial traits and the substantial online presence of Filipinos. Mok and Ren opined that integrated STEM education is the key to helping students develop innovation and entrepreneurial spirit that is required for success in the 21st-century marketplace. For Bhutan, STEM education appears to hold the key to improving the quality of education in Bhutan by increasing the levels of critical thinking, communication, and problem-solving skills among Bhutanese students (Sherab et al.).

Design of STEM curriculum

Engagement in inquiry

Meaningful engagement in the inquiry process of STEM learning, particularly in deciding what and how to measure, observe, and sample; how to design investigations to collect data; how to document and record results in a systematic manner, make decisions about which data is useful in crafting explanations; and how they can be represented, are important for learners (Duschl & Bybee, 2014). As part of their effort to move towards an integrated STEM approach in Bhutan, Sherab et al. applied the 5E inquiry learning cycle to plan learning experiences around the topic of gene expression. The inquiry learning experiences are student- and activity-centric to create opportunities for students to ask questions, investigate, and craft explanations to learn concepts related to gene expression. Their study found that students who are engaged with inquiry have an increased level of interest, understanding, satisfaction, and perceived the concept of gene expression to be less difficult. This finding affirms the value of the inquiry process in learning STEM subjects. Similarly, Mok and Ren described the maker STEM enrichment programme that has strong inquiry characteristics by getting students to engage in generative thinking of solutions, encouraging design of hands-on experimentation, trialling, and fine-tuning their ideas to apply the best idea to devise a problem-solving model. The tenets of inquiry such as raising questions, decision making based on data, and representing evidence are important and useful in integrated STEM problem solving. Problem solving in integrated STEM should take place in the context where sound scientific, mathematical, and technological conceptual knowledge is applied.

Measuring learning outcomes

Singapore, the Philippines, Hong Kong, and Bhutan are all Asian economies in different phases of economic growth. As such, while the authors shared similar reasons (support economic growth and increasing innovation) why their respective economies are embarking on their respective STEM journeys, there are different levels of readiness, contextual, and social factors that shape the emphasis that each education system is giving to their STEM education initiative. The two countries in ASEAN, Singapore and the Philippines, are harnessing their strengths in widespread basic STEM literacies to engage students, teachers, industries and the community in their STEM efforts. From the Singapore case example, priority was given to understanding students' learning experience when planning a STEM curriculum. To illustrate how students' STEM learning experiences can be evaluated, Toh et al. used the idea of STEM capital to assess the effectiveness of a specific STEM curriculum in Singapore. The constructs contributing to STEM capital are (1) Students' views about STEM lessons, (2) Students' attitude towards STEM, (3) Self-concept in learning STEM, (4) Construction of STEM identities, and (5) Career decisions in STEM. These five constructs are centred around the learner and are appropriate in evaluating the effectiveness of any STEM curriculum. Similarly, Paderna and Monterola (this volume) evaluated their STEM curriculum in the Philippines using pre- and post-tests in the areas of (1) STEM career readiness, (2) Metacognitive awareness, (3) Grit, (4) Growth mindset, and (5) Resilience. Compared with STEM capital described by Toh et al. these five constructs are less specific to STEM and are more generic to learning. Their findings suggested the multitude of factors (such as industrial engagement, readiness of teachers, developing a culture of innovation and collaboration, etc.) that need to be considered when designing and planning STEM curriculum. They reported that while students can have positive attitudes towards STEM and appreciate the necessity of STEM in the society, this does not guarantee that the students are interested active problem solvers who will actively pursue an interest in STEM. The authors from both Chapters 6 and 7 relied on feedback by students to fine-tune and re-design their STEM curriculum. The focus on students' feedback is one that is important since the learners are the ones who experienced the curriculum. However, another aspect that is equally important for STEM curriculum refinement is the voices of teachers as well as external stakeholders. Teachers need to be competent and comfortable to implement the STEM programmes. The stakeholders can provide valuable inputs to the relevance of the materials forming the STEM program.

Beyond ASEAN, in the larger Asia, scholars from Hong Kong and Bhutan also shared two case examples of how school-based efforts incorporating maker education and scientific inquiry, were made by teachers and students to shift from monodisciplinary STEM learning to integrated STEM problem solving. Mok and Ren in Hong Kong and Sherab et al. in Bhutan detailed how they collected evidence such as students' responses to pre-/post-tests and analysis of students' artefacts to show that engaging students in more inquiry, hands-on experimentation leads to greater conceptual understanding of science and mathematics.

Partnerships to enhance STEM curriculum

The STEM programme highlighted by Paderna and Monterola is steeped in collaboration with industries and the community. The engagements with various stakeholders within the community allow for authentic STEM ideas and innovations to make their way into teacher professional development and school curriculum. Their curriculum design included inquiry, visits to industries, and networking, including networking with experts in the specific STEM fields. Similarly, the applied STEM programmes described by Toh et al. have different tiers of involvement with the highest tiers involving external partners such as universities and industries. Working in teams is an important feature, particularly teaming with industrial partners since they have the latest technical know-how. Pinnell et al. (2013) reported the importance of university, industry partnership in STEM curriculum design and uplifting science teachers' STEM competencies. They highlighted the importance of having a shared vision of STEM education to help stakeholders get involved in moving STEM education forward.

Challenges and way forward

Evidence from the four chapters highlighted an important aspect in STEM curriculum design and planning – STEM curriculum design needs to be systematic and be driven by clear goals beyond acquisition of conceptual knowledge. A more holistic, pervasive, multi-level way of planning and evaluating STEM learning, such as the five constructs of STEM capital could help to move the design of STEM curriculum forward. The infusion of the three key areas of inquiry process, evaluation of student's learning experiences and partnerships with industries are common and important features for integrated STEM curriculum. There are some challenges to overcome in the creation of meaningful integrated STEM curriculum.

First, teacher competencies and readiness to engage in inquiry as well as STEM problem solving need to be considered. Laforce et al. (2016) proposed 14 foundations for staff support in inclusive STEM classrooms which are also applicable to other STEM classrooms. Some of these foundation supports include engaging staff in professional development or growth activities, freeing up time for staff to reflect on their work, creating time and space for group planning as well as individual planning, providing structures to support risk-taking, and giving staff autonomy. In developing integrated STEM curriculum and implementing

STEM lessons, teachers must be supported to play multiple roles such as a learner, risk-taker, inquirer, curriculum designer, negotiator, collaborator, and teacher (Slavit et al., 2016). In this respect, a research question that is worthy of exploration is *What is the nature of STEM leadership that will develop and support teachers of STEM to design and implement meaningful STEM curriculum?*

Second, students' competencies and readiness need to be considered when designing integrated STEM activities. Students need to be guided to make meaningful connections within and between disciplines of science, mathematics, technology, and engineering. Dierdorp et al. (2014) advocated for the concept-context approach to help students connect between what is learnt and the real world. Besides making meaningful connections of conceptual knowledge, students need to learn to work with others in a group to plan, execute their plans, and also to evaluate if their plans work (Herro et al., 2017). The students in Singapore, Hong Kong, the Philippines, and Bhutan are used to teacher-centric and highly scaffolded forms of learning. As such, to meaningfully engage these students with integrated STEM learning, time and effort is needed to build up students' competencies in inquiry, argumentation, construction, analysis, and ability to work with others. A research question that scholars might be interested to examine is *What skills and competencies do Asian students need to develop to engage with integrated STEM problem solving?*

Lastly, with respect to collaboration with schools, industries, and the community, there is little known about how partnership models can be sustained, scalable, meaningful, and agentic for all parties involved. Industries possess a wealth of knowledge and expertise that would be beneficial for both teachers as well as students as they make connections between the STEM concepts in school to real-world experiences. Collaboration with industries and the community require huge investment of time and energy. Simply attending a one-off workshop or listening to an hour of lecture by an expert is insufficient for sustained interest and learning. Further, it is often difficult to coordinate between school schedules and that of the community or the industries. The direct impact of school-industry-community is also an area that requires more attention to justify the vast amount of resources that goes into building these partnerships. STEM education researchers could possibly explore questions such as, *What are the factors that affect the partnership between schools, industries, and the community? What partnership model would result in mutual benefits for all stakeholders? What are ways to evaluate the impact of school-industry-community partnership in STEM education? What leadership qualities and style would promote STEM cultures in schools?*

References

Dierdorp, A., Bakker, A., van Maanen, J. A., & Eijkelhof, H. M. C. (2014). Meaningful statistics in professional practices as a bridge between mathematics and science: An evaluation of a design research project. *International Journal of STEM Education*, *1*(9), 1–15.

Duschl, R. A., & Bybee, R. W. (2014). Planning and carrying out investigations: An entry to learning and to teacher professional development around NGSS science and engineering practices. *International Journal of STEM Education*, *1*(12), 1–9.

Herro, D., Quigley, C., Andrews, J., & Delacruz, G. (2017). Co-measure: Developing an assessment for student collaboration in STEAM activities. *International Journal of STEM Education*, *4*(26), 1–12, doi: 10.1186/s40594-017-0094-z.

Honey, M., Pearson, G., & Schweingruber, H. (2014). *STEM integration in K-12 education: Status, prospects and an agenda for research.* Washington, DC: The National Academies Press.

Laforce, M., Noble, E., King, H., Century, J., Blackwell, C., Holt, S., Ibrahim, A., & Loo, S. (2016). The eight essential elements of inclusive STEM high schools. *International Journal of STEM Education*, *3*(12), 1–11.

Pinnell, M., Rowley, J., Preiss, S., Blust, R. P., & Beach, R. (2013). Bridging the gap between engineering design and PK-12 curriculum development through the use of the STEM Education Quality Framework. *Journal of STEM Education*, *14*(4), 28–34.

Schleicher, A. (2019). *PISA 2018 – Insights and interpretations.* OECD. Downloaded on May 17, 2021 from https://www.oecd.org/pisa/PISA%202018%20Insights%20and%20Interpretations%20FINAL%20PDF.pdf

Slavit, D., Nelson, T. H., & Lesseig, K. (2016). The teachers' role in developing, opening and nurturing inclusive STEM-focused school. *International Journal of STEM Education*, *3*(7), 1–17.

6 The contextual and personal aspects of the development of a school-based STEM curriculum

A case in Hong Kong

Ida Ah Chee Mok and Zhipeng Ren

Introduction

STEM education, an acronym for Science, Technology, Engineering, and Mathematics, has become a forefront educational research agenda as a result of the worldwide trend to promote integrated education that better prepares learners to solve increasingly complex real-world issues. The origin of STEM education dates back to the 1990s when the National Science Foundation (NSF) in the United States used the term "SMET" to describe the integration of the four areas. The term STEM was used in Judith Ramaley's personal email for policies on education and human resources at National Science Foundation (NSF) (Chute, 2009; Donahoe, 2013).

The traditional compartmentalisation of STEM disciplines is inadequate to meet the demands of the 21st-century workplace and the development of STEM education seems a possible way to develop the 21st-century workforce. Educators and researchers, in general. agree that STEM learning experiences help develop students' 21st-century skills, such as creative abilities, collaborative skills, problem-solving abilities, communication skills, and application of relevant knowledge (Shernoff et al., 2017). In addition, researchers stress the integration in different fields in STEM education covering all levels including early childhood, primary school, secondary school, higher education, and adult education (Banning & Folkestad, 2012). Literature shows that the definition of STEM education and pedagogical approaches are not necessarily uniform and there may be variation between projects and between contexts (English, 2016). For some, STEM education refers to education in science, technology, engineering, and mathematics, while social sciences and other related fields are excluded. There are also studies that integrate STEM into social studies, for example, history classes in middle schools (Pryor et al., 2016). For broader integration of other disciplines with STEM, the acronym STEAM is used in some contexts, where "A" represents the arts – humanities, language arts, dance, drama, music, visual arts, design, and new media (Wade-Leeuwen et al., 2018). Nonetheless, the interdisciplinary and multidisciplinary approaches advocated in STEM education pose a great challenge in the teacher development for STEM education for the demand in pedagogical content knowledge, matching of students' interest

DOI: 10.4324/9781003099888-6

and capacity, assessment for students' learning outcomes is met with a perceived lack of resources (Lee et al. 2019). The teaching and learning of STEM hence pose great challenges to school curriculum, policy as well as teachers.

There is a dearth of research about STEM education in Asia. For example, Lee et al. (2019) reported that out of the 662 published papers on the Web of Science, only 8.5% of the papers were based on studies carried out in Asia. As such, more work is needed to illumine the STEM efforts in Asia. This chapter attempts to make a contribution to our understanding of STEM curriculum efforts in Asia by exploring the contextual and personal aspects of the development of a school-based STEM curriculum in Hong Kong.

We have organised this chapter into four parts: (1) Challenges for teachers in STEM teaching and learning, (2) An overview of the STEM education policy for Hong Kong, (3) An example of the development of a primary school-based STEM curriculum, and (4) Exploring the contextual and personal aspects of the development of a school-based STEM curriculum.

Challenges in STEM teaching and learning

This section reviews the literature related to the challenges for teachers in STEM teaching and learning, within an Asian context.

School district support and administrative guidance could be an important factor for STEM implementation in Asia. In the study by Park et al. (2016), they found that teachers in South Korea felt that administrative and financial support could be a challenge to STEM implementation. The East Asian culture in South Korea may play a certain role in the administrative factors in the implementation of STEM education, for example, how to handle large class sizes, the competitive education system, the availability of resources for individual schools, and the matching of students' interest and ability. In Park's study of teachers' perceptions in South Korea, the challenges include lacking administrative and financial support, lacking time to prepare STEAM lessons, increased workload, and difficulties in using media and experimental equipment. The demand on resources such as time, teachers' development, and facilities may be a major resistance to changes in pedagogical approaches. In addition, other factors such as competitive examination systems, parental expectations, are very different from other places outside the East Asian regions such as America, Europe, and United Kingdom. Such cultural phenomenon still needs further research to examine the difficulties and challenges they impose in the implementation of STEM education.

Teachers in many economies may also face various difficulties and challenges in STEM teaching and curriculum design (Margot & Kettler, 2019). The mathematics teachers in the study of Thibaut et al. were found to face more difficulties in implementing STEM courses, mainly due to perceived challenges that mathematics teachers have to help students to solve authentic and real-world problems (Thibaut et al., 2018a). In STEM classrooms that engage in problem solving, STEM teachers have to maintain the class order, guide students and take on a non-authoritative role to facilitate and support students in the co-construction of knowledge and skills. In addition to the challenges mentioned earlier, the teaching and learning of many STEM topics may have technological elements (e.g., robots, use of IT resources, statistical tools). Hence, to better implement STEM education, STEM teacher development in many aspects is important. For example, workshops for STEM teachers can be held (Chai, 2019) to communicate technical information and gain peer support.

In the study of Park et al. (2016) in South Korea, teachers complained that they suffered increased workload associated with STEM programming when preparing teaching materials and allowing for varying ability levels among students. Teachers from Saudi Arabia felt the collaboration and interdisciplinarity could be difficult for them to grasp and understand, especially how to integrate four STEM disciplines together and how to integrate technology (Heba et al., 2017). In the study of teachers' self-efficacy and concerns for STEM education by Geng et al., the findings showed that only 5% of the teachers regarded themselves as "well prepared" (Geng et al., 2019), suggesting that it is essential to raise teachers' self-efficacy by pedagogic support and professional training.

A study by Fore et al. (2015) explored the factors that influence secondary STEM teacher learning and change, as well as the processes via a case study design of a professional development programme. The results showed that the outcomes of professional development programme are heavily influenced by teachers' perceptions and knowledge, as well as how they internalised via their negotiation with the material constraints present in educational policy and socioeconomic realities. The results of a study of Korean pre-service teacher shows that pre-service teachers' art appreciation, attitude toward science, and technology acceptance had a significant effect on their creative convergence competency, which in turn, have an impact on their teaching competency in STEAM education (So et al., 2019). Dong et al., (2019) carried out a survey of 458 Chinese teachers with results showing that teaching self-efficacy, pedagogical design self-efficacy, and collegial support were the important predictors of teachers' engagement in STEM teaching, whereas, teaching self-efficacy was influenced by teachers' knowledge and administration support.

The aforementioned shows that the teacher is a pertinent element for STEM education, whereas, the factors related to the teacher are, in turn, associated with the contextual factors such as education system and the school environment, resources, and teacher development.

To conclude, the promotion, design, and implementation of STEM education in Asian economies pose challenges and difficulties to curriculum both contextually and personally. In what follows, we will discuss an example of the development of a school-based STEM curriculum in the context of the STEM curriculum policy in Hong Kong. While STEM education policy has been discussed in an earlier chapter by So et al., we will provide a brief account of the policy here with a focus on curriculum policies to set the context for our discussion about how these policies play out in the classroom. Specifically, we distil the contextual and personal aspects of implementation shaped by the STEM curriculum policy. The contextual aspects refer to curriculum policy and administration in the system and in the school, whereas, the personal aspects refer to the factors and concerns related to the teachers who are the key players in the curriculum reform.

An overview of the STEM curriculum policy for Hong Kong

Hong Kong, as a special administrative region of China, is currently promoting STEM education as a critical step towards fostering an innovative mindset among students (Hong Kong Education Bureau [HKEDB], 2016). The HKEDB (2016) proposed a report on the promotion of STEM education and regarded STEM education as a key emphasis in the ongoing renewal of the mathematics and science curriculum. These efforts aimed to strengthen "Hong Kong students' abilities to integrate and apply knowledge and skills across different STEM disciplines, and to nurture their creativity, collaboration and problem-solving skills, as well as to foster their innovation and entrepreneurial spirit as required in the 21st century". Considering the rapid development of scientific, technological, mathematical, and engineering knowledge in the present and future, STEM education is positively encouraged. For a background of this chapter, the curriculum policy of STEM promotion in primary and secondary schools in Hong Kong are recapitulated below.

1 *Renewal of key learning areas in the curricula of science, technology, and mathematics.* The updated version of Key Learning Areas (KLAs) Curriculum Guides of Science, Technology, and Mathematics Education was published in 2017 by HKEDB. This renewal of curriculum documents has integrated suggestions for the direction of STEM education with the existing curriculum aims and frameworks, serving as references for school teachers to design their school-based STEM teaching and learning activities, materials, assessment, and so on. Recommendations are also given in the three revised curriculum guides about the interdisciplinary activities, cross-KLAs collaboration, and learner diversity. For example, two approaches to organising learning activities on STEM education are proposed in both mathematics and science education KLA curriculum guides: "learning activities based on topics of a KLA for students to integrate relevant learning elements from other KLAs"; "Projects for students to integrate relevant learning environments from different KLAs" (HKEDB, 2017a, 2017b). Technology KLA also contributes to the promotion of STEM education through developing a solid knowledge base; strengthening students' ability to integrate and apply knowledge and skills; fostering innovation in meeting the challenges of economic and technological development; and strengthening the collaboration among teachers and partnerships (HKEDB, 2017c).

2 *Enrichment of learning activities for students.* The purpose of enriching student learning activities is "to organise an education fair for students to showcase and celebrate a wide range of student achievements on STEM-related

areas on a regular basis". Evidence has shown the positive relationship between the efficiency of learning activities and learning achievements in STEM education (Chiu et al., 2015; Kakarndee et al., 2018). The policy suggestions for primary and junior secondary schools are to provide more opportunities inside and outside the classroom, for students to engage in STEM-related learning activities. The school-based support for holistic curriculum planning and implementation is strengthened via collaboration with other institutions, like The Academy of Hong Kong, Hong Kong Science and Technology Parks Corporation, to increase the provision of various STEM learning opportunities for all students.

In summary, the promotion of STEM education in Hong Kong is not the same as other local curriculum reforms in the past, which stipulated the change of contents for a subject discipline. The policy, in this case, suggests an overarching direction, strategies, and resources for schools to craft their own school-based curriculum. School principals and teachers have supported the policy as it provides lots of opportunities and freedom for the schools to explore, they own tailor-made curriculum and share in the education professional community.

An example of the development of a primary school-based STEM curriculum

In line with the government's promotion of STEM education, schools were encouraged to develop their school-based STEM curriculum. However, curriculum reforms do not always report success for the factors purporting the successful cases might be complex and much depend on the individual school and teachers. In this section, via the report of the experience of a successful case in a Hong Kong primary school, we attempt to provide some insights into the processes of developing a school-based STEM curriculum while capitalising on the school's mathematics enrichment project in the backdrop.

The case study school

The school is a very popular primary school in the neighbourhood, managed by a registered non-profit organisation with a mission to promote the development of quality education in Hong Kong through setting up quality schools. The school implemented an ability-grouping policy with a selection strategy based on the results of the formative and summative assessment, the teachers' observation, supplemented with the guidelines in the *Education Bureau Web-based Learning Courses* (Hong Kong Education Bureau, 2013). The school policy also facilitated a good teaching environment from the teachers' perspective, such as appropriate grouping of the students and appointment of the teachers with relevant expertise for the programme. The teachers would also teach more than one class for the same year level to enable them the opportunity in a relatively short time to improve and modify their own lessons based on self-reflection and catering for student diversity.

For the research purpose of this study, a letter was written to the school principal for access to the school's data of the implementation of the enrichment programme and the STEM programme, which includes: Minutes of school meetings, PowerPoint presentations of different occasions, dissemination seminar, lesson materials such as lesson plans, worksheets, lesson video clips, and student work, a presentation of the internal school survey, and the key teacher's personal audio record of the PowerPoint presentation.

The development of the school-based STEM curriculum was assigned to a key-teacher, Mr Lee (pseudonym), as he was the lead teacher of the school's enrichment programme, which had been running from 2010 to 2015. The new school STEM curriculum was an extension of the school's enrichment programme from 2015 in alignment with the government's promotion of STEM education. Mr Lee was an enthusiastic and passionate mathematics teacher, had a Master of Education degree specialised in mathematics education, was a core member of the Gifted Education Network (Mathematics) in the Education Bureau, and an awardee of the Certificate of Teaching Excellence Award 2016. Mr Lee designed and implemented the enrichment programme that adopted the problem-solving approach (Polya, 2014).

Based on the Trends in International Mathematics and Science Study (TIMSS), Hong Kong students scored relatively well in mathematics as compared to students in other parts of the world. However, their confidence in mathematics and engagement in mathematics learning was relatively lower than in other Asian regions (Mullis et al., 2020). Also, the teachers in the school reported that the regular curriculum did not serve the purpose of exerting the potential of high ability students (Mullis et al., 2020). Hence, the overall aim of the enrichment programme was to raise students' learning interest and confidence in mathematics through enrichment of their learning of mathematics to enhance their thinking skills and problem-solving skills. The programme started with 30 Primary 5 students that belonged to the top 25% of the class based on their Primary 3 performance in 2011/12. It extended progressively to include three levels (Primary 4, 5, and 6) progressively in 2012/13, then integrated with STEM reform initiatives in 2015.

The specific objectives for the school's implementation of the enrichment programme were: (1) To provide sufficient opportunities for students to develop their communication, analysis, and problem-solving skills; and (2) To let students fully utilise their normal lesson time. The reformed school-based mathematics gifted curriculum was a three-year enrichment programme for gifted students in Primary 4 to Primary 6.

The school evaluated the students' academic performance before and after the enrichment programme. A total of 90 students – 15 students from the high ability group and 15 students from the middle ability groups in each of Primary 4, 5, and 6 classes – participated in the evaluation of the impact of the programme

(Sung, 2017). According to the report by Sung (2017), the following hypotheses were made:

> H1: The programme helped the students in the high ability group get higher academic achievements.
>
> H2: The programme helped the students in the high ability group develop high order thinking ability.
>
> H3: The self-esteem of the students in the high ability group was raised.
>
> H4: The test anxiety of the students in the high ability group was higher.

The assessment results included formative and summative assessments. The test for H2 was based on the formative assessment of non-conventional questions. The tests for H3, H4 were based on a questionnaire designed to collect the information of non-cognitive measurement for attitudes, beliefs, and values, supplemented with student interviews. The results (Table 6.1) showed that Primary 6 students benefited most from the programme and that the enrichment programme could be a double-edge blade for some Primary 4 students. Possible reasons were that the students might take time to get used to the programme and the students were likely to benefit more after having participated in the programme for a longer period. In addition, the school graduates sent back in retrospect very positive feedback about how they benefitted from the programme.

From the teachers' perspectives and experiences, there were four major strategies for implementing the programme, namely, (1) Compact and fast learning pace; (2) Enrichment topics for a board knowledge base and interest; (3) Extension of existing topics for developing advanced thinking skills; and (4) Realistic contexts. These features were aligned to the rationales of STEM education for the purpose of developing students' 21st-century skills. Thus, it was feasible to integrate STEM as an extension for the enrichment programme in the school curriculum. For the design of the STEM programme, the teachers adapted the Maker concept with a belief that the nurturing of Makers should be that of helping students become good problem solvers, thus going beyond the experiences of assembling a toy from parts. To achieve this aim, the school developed a school-based strategy for the STEM programme with four features:

Table 6.1 The summary of the tests for the hypotheses

Hypothesis	P.4 High ability group	P.5 High ability group	P.6 High ability group
Academic achievement	Negative (not significant)	–	Positive
High order thinking	–	Positive	Very positive
Self-esteem	–	–	Positive
Test anxiety	Positive	–	–

1 Creation of "Thinking Space": Some aspects of promoting the students' thinking opportunity included: The students' ideas were valued and they were encouraged to find out more possible solutions, alternative solutions, and "proof" or justification for what they had learned.

2 Creation of Hands-on Experience: The traditional mathematics curriculum provided little opportunity for hands-on experiences; hence, the students' ideas often became wishes that never had a chance to be put to the test. In this model, the students were encouraged to put their ideas into hands-on experience.

3 Provision of Trial-and-Error Experience: In the encouragement of experimenting with their ideas, students also developed a deep learning for what were fair experiments and systematic trials.

4 Applying Polya's Problem-Solving Model: Based on the guidelines for systematic trials, Polya's problem-solving model was applied to reflect upon the trials for identifying the crux of the problems and means for solving the problem.

In designing the STEM lessons, gender issues were considered. According to the teacher, "Though some girls may not like mathematics, while choosing the topics carefully, the girls also develop in an interest in mathematics". In other words, the topics could be chosen while taking the girls' interests into consideration. Some examples of the enrichment topics were modified to STEM topics, such as, building a shooting platform for paper cannonballs, building a maglev toy car, making a musical instrument, and handling Big Data.

An example: The topic of "sound"

The topic of "sound" was first created and tried on the high-ability group in Primary 6 in a previous enrichment programme. It was subsequently revised and implemented for all the Primary 6 students in E100STEM (course code). The investigation of "sound" was a major part of E100STEM as it lasted for about 12 to 15 lessons. One lesson was conducted every Friday, starting in May or early June, after the class had finished the regular mathematics and the old enrichment mathematics component. This arrangement was imperative for it provided curriculum time for the implementation.

The design of the learning process embraced the elements of enquiry and problem solving. In addition to knowing the facts about the transmission and frequency of sound, the students learned to raise many questions, and they had to design their own experiments and methods to find out the answers. Some questions included: "Is sound really transmitted by air?", "Besides the vacuum test, are there other methods?", "Can we find methods to measure the speed of sound?", "How may the water in the glass affect the sound produced?", "Is there a formula?", "How true is the statement 'the range of human hearing is 20Hz to 20,000Hz'?". These questions showed that the students had thought about the experiments in-depth, with enthusiasm, and critically, based on their own knowledge of mathematics and science, and were in quest of further extension. The students designed tests and collected data from different people including members of their families. Then they discussed the trends of the processed data such as line charts and drew conclusions from the findings.

Finally, they embarked on a project to make an instrument and play a song with their self-made instrument. The students applied their mathematics concepts and other knowledge about sound to find answers to their questions about the frequency and transmission of sound. To make the musical instrument, they had to decide how to cut the long brass pipe into pieces of different lengths based on their own experiments. If they had made unwarranted guesses, there would be a lot of waste of materials and frustration. Eventually, they solved the problem by drawing a line of best fit to find the relationship between the length of the brass pipe and the musical note. They presented their ideas confidently to the class during the project presentation. In the students' written reports, they explained the principles they applied in the methods, the difficulties they encountered in the process, and how they solved the problems. In addition, they played their self-made musical instrument in the final presentation in front of the whole class.

Exploring the contextual and personal aspects of the development of a school-based STEM curriculum

The school-based STEM curriculum reported in this chapter took place in a primary school in Hong Kong where the traditional disciplinary curriculum might not necessarily afford an adequate environment to equip the students with 21st-century skills and capacity, without strong government support. Very often, there is a gap between intended and implemented curriculum, for the attention will shift to the monitoring of outcomes as soon as an innovation is planned (Fullan & Pomfret, 1977). Successful implementation often depends on both contextual factors and personal factors. To address these challenges, having a top-down policy can play an important role in pushing for education change as the impetus for school-based changes is achieving a closer alignment with the direction of curriculum reforms advocated by the Hong Kong Education Bureau. The example discussed in this chapter took place in Hong Kong where students have demonstrated good performance in international comparative studies such as TIMSS (Mullis et al., 2020) and Programme for International Student Assessment (PISA) (OECD, 2021). These reports may, at times, serve as a double-edged sword. On the one hand, the reports suggested that the students had achieved a strong foundational knowledge in STEM subjects such as mathematics and science. On the other hand, the reports raise questions on whether any education change will benefit students learning outcomes. Furthermore, the top-down directives may also change the culture outside schools. Like other places in the world, the need for strengthening teachers STEM pedagogical knowledge (Yıldırım & Şahin Topalcengiz, 2019), is a great challenge in Hong Kong. The policy had provided a lot of opportunities for interdisciplinary sharing and partnerships outside school, hence shedding insights on teacher professional development (see HKEDB, 2016; Lin, 2019; Yeo, 2020). In addition, as STEM is not an independent subject in the curriculum, there is no prescribed content, curriculum development within the school may take on diverse forms. However, the success of the school-based curriculum does not happen by chance. The environment outside and inside the school is important. While Mr Lee developed the STEM programme for his school, he capitalised on the strengths (pedagogy with enhanced problem-solving, enquiry and collaborative elements, STEM-related topics, evidence of students' achievements) of an ongoing enrichment programme in the school curriculum to push for STEM learning.

Another factor for successful curriculum reform is teachers (Craig, 2006; Parker, 2015; Salokangas, et al., 2020). Mr Lee, who played the role of the key teacher, was an enthusiastic and knowledgeable teacher. He had rich teaching experience for primary mathematics, a master degree in mathematics education, and had been invited to give many workshops about his work on student enrichment and STEM curriculum in both local and overseas conferences. His devotion and passion in his workplace were certainly indispensable in the process of the curriculum design and implementation. Yet, the context of school-based curriculum reform also provided an important ground for his professional agency (Priestley et al., 2015), building upon his roles and responsibilities in the process. In other words, a teacher's professional achievement of agency can be described in terms of how the teacher interacted with people and resources in his working place and the larger community context (Priestley et al., 2015). Curriculum changes would inevitably involve many stakeholders, namely, school management boards, principals, teachers, students, and parents. One of the teachers' concerns for STEM education is about how students may benefit in the process (Geng et al., 2019). In addition, having holistic student evaluation that includes academic achievement, high-order thinking, self-esteem, and test anxiety, is important. Such an evidence-based approach not only provides evaluative feedback for the programme but also supports teachers to reflect, provide evidence for reporting to various stakeholders, and sustain in the development process. Mr Lee's work had provided room for further developing their potential and obtaining positive results in academic achievement, higher-order thinking, and self-esteem, and releasing test stress. The students were also aware that their experience has developed their potential and some graduates of the school expressed special gratitude for what they have learned and achieved in these school-based programmes. Mr Lee, the key teacher in the school-based reform, designed the lessons with a genuine consideration of the curriculum goals and careful application of relevant learning theories in his design of the lessons, catering for the students' capacities. The problem-solving, enquiry, and collaborative elements in his pedagogical approaches, and evidence-based evaluation of student's academic achievement and attitude, has helped different stakeholders including students, parents, colleagues, and the school management board; and these are the crux to substantiate the innovative elements in school-based curriculum. Furthermore, the plan was implemented, and the lessons were scrutinised and evaluated with a deep reflection of the student's achievement in both cognitive and affective domains.

Conclusion

The teacher is the key person in the implementation of curriculum. In addition to the teachers' personal capacity, teachers can only work within various constraints of the system that includes policy, resources, concerns for school vision, parents, colleagues, and students. This chapter examined the personal and contextual aspects of the development of a school-based curriculum STEM via a case in the context of the Hong Kong curriculum STEM policy. While this chapter shows that top-down policy change can push for change in the classroom, other factors also come into the picture. Firstly, dedicating the leadership role to a teacher with passion and giving room for the teacher to exercise professional capacity. Mr Lee in this case is a devoted teacher who continuously develops his agency (Priestley et al., 2015) via academic study, teaching design to enhance students' development, self-reflection, and is generous in sharing with local teachers and professionals. The implementation of the innovative school-based programme evaluation shows that the students had positive results in academic achievement, higher-order thinking, and self-esteem. Such evidence-based approach has made the outcomes transparent to school management board, teachers, students, and parents, and can indirectly help appreciate the change and the teachers' worthy effort, hence giving a strong rationale for sustainability and continuous development.

References

Al Salami, M. K., Makela, C. J., & De Miranda, M. A. (2017). Assessing changes in teachers' attitudes toward interdisciplinary STEM teaching. *International Journal of Technology and Design Education*, 27(1), 63–88.

Banning, J., & Folkestad, J. E. (2012). STEM education related dissertation abstracts: A bounded qualitative meta-study. *Journal of Science Education and Technology*, 21(6), 730–741.

Chai, C. S. (2019). Teacher professional development for science, technology, engineering and mathematics (STEM) education: A review from the perspectives of technological pedagogical content (TPACK). *The Asia-Pacific Education Researcher*, 28(1), 5–13.

Chiu, A., Price, C. A., & Ovrahim, E. (2015). *Supporting Elementary and Middle School STEM Education at the Whole School Level: A Review of the Literature*. In NARST 2015 Annual Conference, Chicago, IL. https://www.msichicago.org/fileadmin/assets/educators/science_leadership_initiative/SLI_Lit_Review.pdf.

Chute, E. (2009). *STEM Education Is Branching Out: Focus Shifts from Making Science, Math Accessible to More Than Just Brightest*. Pittsburg Post-Gazette. http://www.post-gazette.com/news/education/2009/02/10/STEMeducation-isbranchingout/stories/200902100165.

Craig, C. (2006). Why is dissemination so difficult? The nature of teacher knowledge and the spread of curriculum reform. *American Educational Research Journal*, 43(2), 257–293.

Donahoe, D. (2013). *The Definition of STEM?* Retrieved 20 April 2021 from: https://insight.ieeeusa.org/articles/the-definition-of-stem/.

Dong, Y., Xu, C., Song, X., Fu, Q., Chai, C. S., & Huang, Y. (2019). Exploring the effects of contextual factors on in-service teachers' engagement in STEM teaching. *The Asia-Pacific Education Researcher*, *28*(1), 25–34.

English, L. D. (2016). STEM education K-12: Perspectives on integration. *International Journal of STEM Education*, *3*(1), 3.

Fore, G. A., Feldhaus, C. R., Sorge, B. H., Agarwal, M., & Varahramyan, K. (2015). Learning at the nano-level: Accounting for complexity in the internalization of secondary STEM teacher professional development. *Teaching and Teacher Education*, *51*, 101–112.

Fullan, M., & Pomfret, A. (1977). Research on curriculum and instruction implementation. *Review of Educational Research*, *47*(2), 335–397.

Geng, J., Jong, M. S. Y., & Chai, C. S. (2019). Hong Kong teachers' self-efficacy and concerns about STEM education. *Asia-Pacific Education Researcher*, *28*(1), 35–45.

Heba, E. D., Mansour, N., Alzaghibi, M., & Alhammad, K. (2017). Context of STEM integration in schools: Views from in-service science teachers. *Eurasia Journal of Mathematics, Science and Technology Education*, *13*(6), 2459–2484.

Hong Kong Education Bureau. (2013). *Education Bureau Web-based Learning Courses.* Retrieved 20 April 2021 from: https://hkage.org.hk/webbasedlearning/learning-course/en/index.php.

Hong Kong Education Bureau (2016). *Promotion of STEM Education: Unleashing Potential in Innovation.* Retrieved 20 April 2021 from: https://www.edb.gov.hk/attachment/tc/curriculum-development/renewal/STEM/STEM%20Overview_c.pdf

Hong Kong Education Bureau. (2017a). *Mathematics Education Key Learning Area Curriculum Guide (Primary 1 – Secondary 6).* Retrieved 20 April 2021 from: https://www.edb.gov.hk/attachment/en/curriculum-development/kla/ma/curr/ME_KLACG_eng_2017_12_08.pdf

Hong Kong Education Bureau. (2017b). *Science Education Key Learning Area Curriculum Guide (Primary 1 – Secondary 6).* Retrieved 20 April 2021 from: https://www.edb.gov.hk/attachment/en/curriculum-development/kla/science-edu/SEKLACG_ENG_2017.pdf

Hong Kong Education Bureau. (2017c). *Technology Education Key Learning Area Curriculum Guide (Primary 1 – Secondary 6).* Retrieved 20 April 2021 from: https://www.edb.gov.hk/attachment/en/curriculum-development/kla/technology-edu/curriculum-doc/TE_KLACG_Eng_5_Dec_2017_r2.pdf

Kakarndee, N., Kudthalang, N., & Jansawang, N. (2018). The integrated learning management using the STEM education for improve learning achievement and creativity in the topic of force and motion at the 9th grade level. AIP Conference Proceedings. doi: https://doi.org/10.1063/1.5019515.

Lee, M. H., Chai, C. S., & Hong, H. Y. (2019). STEM education in Asia Pacific: Challenges and development. *The Asia-Pacific Education Researcher*, *28*, 1–4.

Margot, K. C., & Kettler, T. (2019). Teachers' perception of STEM integration and education: A systematic literature review. *International Journal of Stem Education*, *6*(1), 1–16.

Mullis, I. V. S., Martin, M. O., Foy, P., Kelly, D. L., & Fishbein, B. (2020). *TIMSS 2019 International Results in Mathematics and Science.* Retrieved from Boston College, TIMSS & PIRLS International Study Center. https://timssandpirls.bc.edu/timss2019/international-results/

OECD. (2021). *Mathematics Performance (PISA) (Indicator)*. doi: 10.1787/04711c74-en (Accessed on 12 May 2021).

Park, H., Byun, S. Y., Sim, J., Han, H. S., & Baek, Y. S. (2016). Teachers' perceptions and practices of STEAM education in South Korea. *Eurasia Journal of Mathematics, Science and Technology Education, 12*(7), 1739–1753.

Parker, G. (2015). Postmodernist perceptions of teacher professionalism: A critique. *The Curriculum Journal, 26*(3), 452–467.

Polya, G. (2014). *How To Solve It: A New Aspect of Mathematical Method*. Princeton University Press: New Jersey, USA.

Priestley, M., Biesta, G., & Robinson, S. (2015). *Teacher Agency: An Ecological Approach*. London; New York: Bloomsbury Academic.

Pryor, B. W., Pryor, C. R., & Kang, R. (2016). Teachers' thoughts on integrating STEM into social studies instruction: Beliefs, attitudes, and behavioral decisions. *The Journal of Social Studies Research, 40*(2), 123–136.

Salokangas, M., Wermke, W., & Harvey, G. (2020). Teachers' autonomy deconstructed: Irish and Finnish teachers' perceptions of decision-making and control. *European Educational Research Journal, 19*(4), 329–350.

Shernoff, D. J., Sinha, S., Bressler, D. M., & Ginsburg, L. (2017). Assessing teacher education and professional development needs for the implementation of integrated approaches to STEM education. *International Journal of Stem Education, 4*(1), 13.

So, H. J., Ryoo, D., Park, H., & Choi, H. (2019). What constitutes Korean pre-service teachers' competency in STEAM education: Examining the multi-functional structure. *The Asia-Pacific Education Researcher, 28*(1), 47–61.

Sung, L. P. W. (2017). The learning impact of high ability students under ability grouping in mathematics. *HKUGA Primary School Collection of Academic Papers: We Teach, We Learn, Learning for the Future* (pp. 28–34).

Thibaut, L., Knipprath, H., Dehaene, W., & Depaepe, F. (2018a). How school context and personal factors relate to teachers' attitudes toward teaching integrated STEM. *International Journal of Technology and Design Education, 28*(3), 631–651.

Wade-Leeuwen, B., Jessica Vovers, J., & Silk, M. (2018). *Explainer: What's the Difference between STEM and STEAM?* June 11, 2018 6.05am AEST https://theconversation.com/explainer-whats-the-difference-between-stem-and-steam-95713.

Yeo, R. (2020). *Hong Kong Organisation Pushing STEM Learning for All is among This Year's Operation Santa Claus beneficiaries*. South China Morning Post, November 6, 2020. https://www.scmp.com/news/hong-kong/education/article/3108429/hong-kong-organisation-pushing-stem-learning-all-among

Yıldırım, B., & Şahin Topalcengiz, E. (2019). STEM pedagogical content knowledge scale (STEMPCK): A validity and reliability study. *Journal of STEM Teacher Education, 53*(2), 2.

7 Advancing integrated STEM education in the Philippines through STEM curriculum implementation

Edwehna Elinore S. Paderna and Sheryl Lyn C. Monterola

Introduction

Science, Technology, Engineering, and Mathematics, or STEM, has increasingly been the focus of the school's curriculum over the past ten years specifically in the United States of America and United Kingdom (Banks & Barlex, as cited in Hallström & Schönborn, 2019). This STEM-focused education was extended to and continues to grow in Asian countries where a deliberate integration of engineering design or design thinking in STEM subjects has been observed or where the problems that have been investigated by students have real-world orientations and are multidisciplinary in nature (Lee et al., 2019). In the case of the Philippines, STEM is introduced in the formal school curriculum as one of the strands in the academic track in senior high school (SHS).

In Southeast Asia, the Philippines is next to Indonesia in terms of manpower pool in the region, with 27,216,398 learners during the school year 2019–2020. Of this enrollment, 3,194,035 are in SHS and only 491,349 (15.38%) are on the STEM strand (Department of Education [DepEd], 2019). Do these numbers indicate students' lack of interest in STEM? Is there an integrated STEM curriculum in the Philippines?

STEM education in the K to 12 senior high school programme

The landscape of basic education in the Philippines changed when the K to 12 Basic Education Curriculum was implemented under the Enhanced Basic Education Act of 2013, a law signed by the former Philippine President Benigno Aquino III (officialgazette.gov.ph, 2013). This curriculum requires Filipinos of this generation to complete 13 years of basic education from Kinder to Grade 12. Prior to the said law, the basic education was only from Kinder to Grade 10.

The SHS Programme (Grades 11 and 12), the new addition to the curriculum, is designed to cater for the needs and interests of students and provide opportunities for possible employment. It allows students to select a track that they intend to pursue. There are four basic tracks: Academic, Technical-Vocational-Livelihood

DOI: 10.4324/9781003099888-7

(TVL), Sports, and Arts and Design. The Academic Track includes the STEM strand, among three others. The STEM strand is intended for students who would be pursuing STEM-related college courses and careers.

The STEM curriculum of the Department of Education includes three subject classifications: *Core subjects, contextualised subjects*, and *specialisation subjects*. STEM subjects such as Earth Science, General Mathematics, and Statistics and Probability fall under the *core subjects* along with other subjects such as Physical Education and Health, Languages, Arts, Philosophy, etc. *Contextualised* subjects are subjects where students can apply what they have learned and what they can utilise for their future academic track and careers. *Contextualised subjects* include Research, Entrepreneurship, Languages for Professional Purpose, and Empowering Technologies. Lastly, the *specialised subjects* include three major fields in science: Biology, Chemistry, and Physics. Each subject is tackled twice for an academic year, with the first subject serving as a pre-requisite to the other. For example, General Chemistry 1 is taken during the first semester and General Chemistry 2 during the second semester. For Mathematics, the specialized subjects are Pre-Calculus for the first semester and Calculus for the second semester (DepEd, 2013).

STEM education programme challenges

According to a number of national and international assessments of student's learning, it has often shown that the state of science and mathematics performance in the Philippines needs improvement. Despite a good number of recognitions brought by Filipinos in the field of science, technology, and innovation, these recognitions are still often overshadowed by the poor standings of Filipinos in terms of science and mathematics education in comparison to other countries (Imam et al., 2014). The rise of multiple reports published about the educational crisis in the past years triggered the dire need for major changes, opportunities, improvements, and expansions in STEM education. Although the call to action is heightened, the transition to and the change in curricula have not been easy (Sahin, 2015).

Among the numerous crises faced by the improvement of STEM education in the country is the lack of educators with exceptional mastery in the field related to STEM (Feliciano et al., 2013). Poor performance in the national Basic Education Exit Assessment (BEEA), which is taken by all exiting SHS Grade 12 students, has been reported. In the 2019 BEEA, the mean percentage score in Science was 36.2% and 27.9% in Mathematics (DepEd, 2019). In terms of the number of students at the tertiary level, only 28.7% of the 3.2 million university and college students are in STEM-related courses such as agriculture, forestry, fisheries, sciences and mathematics, engineering, and information technology (CHED Statistics, 2019). The STEM pipeline further narrows as reflected by the low number of research and development professionals. Specifically, the ratio of research and development personnel per million population is only 245 compared to Malaysia that has 3,912 and Vietnam with 1,825 R&D personnel per million population (UNESCO UIS, 2019).

Globally, STEM education aims to develop STEM engagement literacy under the 21st-century competencies and a workforce that is STEM-capable (Honey et al., 2014). These goals are developed for students to become critical and creative thinkers who can develop innovations, solutions, and sound decisions. The purpose of STEM education is to produce a workforce that is capable of contributing significantly to the industry of their choosing. Similarly, the Philippines shares the same STEM goals, with slight modification, which are anchored to DepEd's four core values – *maka-Diyos (for God), maka-tao (for the people), makakalikasan (for nature),* and *makabansa (for the nation)* (Vera Cruz et al., 2018). Despite these common goals, STEM education does not show significant integration across its subjects.

Integration of STEM subjects is not evident in both basic education and higher education. In particular, at the basic education level, all sciences and mathematics are tackled as different subjects without underlying themes. Looking through the STEM curriculum for SHS, it is clear that STEM subjects are all taught separately. As mentioned, the three major science fields are taught as separate subjects across the semesters from Grade 11 to Grade 12. Additionally, most students treat each STEM discipline as separate subjects and their ability to make connections across these disciplines is difficult to measure (Honey et al., 2014). This also affects their collaborative capabilities in problem solving in which STEM can be applied. Furthermore, every STEM subject is assessed separately through its own standardised achievement test which fails to measure the connections among STEM subjects. For instance, inter-school research competitions are classified into different categories based on a specific science field. Competing students for science research categorise their studies under either life or applied science which is assessed separately. In junior high school, though all three major branches of science are taught in every grade level, there is still a clear demarcation; that is, where they are taught in separate grading periods with unclear underlying themes and integration. The integration of STEM subjects and across other learning disciplines should be targeted. When such is achieved, learners' critical and creative thinking skills will be developed and consequently be applied in innovation, problem solving, and decision making.

Rationale for a STEAM innovation programme

Crucial to nation-building and economic prosperity is a citizenry who are not only consumers or end-users but are also producers, entrepreneurs, creators, and innovators. The development of such citizenry is deeply rooted in human capital development through high-quality education and training that foster a culture of innovation and collaboration.

Filipinos may already have the seeds and fertile ground for which innovative and collaborative mindsets can be cultivated. The 2015 Global Entrepreneurship Monitor reported that the perceived entrepreneurial traits of Filipinos are above world averages and are also higher compared to their Asian neighbours (Sansano, 2018). For instance, Filipinos rated higher in terms of perceived opportunities, perceived capacities, entrepreneurial spirit, and innovation compared to Malaysia, Vietnam, and Thailand. Moreover, Filipinos already have a substantial online presence having 67 million internet users who spend an average of nine hours a day on the internet (We Are Social, 2018).

Despite the high entrepreneurial and innovative spirit of Filipinos, the Philippines' global innovation performance only ranks 73rd in the world, compared to its Asian counterparts such as Malaysia (ranked 35th), Thailand (44th), and Vietnam (45th) (Sansano, 2018). Furthermore, the strong digital connectedness of Filipinos has yet to be channelled into a thriving e-commerce. The Philippines only gets 39% share of people buying online in contrast to Malaysia and Thailand, which have 59% and 62% share, respectively (Sansano, 2018). The way to turn this around is through human capital development – education and training that consciously create a culture of innovation and collaboration. This is strongly articulated in the agenda of various government agencies such as Ambisyon 2040: The Philippine Development Plan (NEDA), Science for the People (DOST), and the Inclusive Innovation Industrial Strategy (i3 Strategy-DTI).

The Department of Education already has a considerable platform for human capital development on innovation and collaboration because of its STEM curriculum in SHS. However, its implementation needs to be supported by sustained capacity-building of both teachers and learners on authentic STEM practices for developing innovative products or optimised processes/services. It also needs a mechanism for sharing students' innovations with their communities. Hence, the STEAM Innovation Programme for SHS has been conceptualised and implemented.

STEAM innovation programme

The STEAM Innovation Programme is a project with Unilab Foundation that aims to create and sustain a culture of innovation and collaboration through partnership with local communities, teacher capacity-building, sustained professional development, and community expos involving public SHS that offer STEM.

Consultation with stakeholders

The first phase of the programme engaged the local community stakeholders. Through a roundtable discussion involving school leaders, local government officials, and local business owners, potential areas for innovation such as local industries that can benefit from research and community problems that can be addressed through scientific means were identified. A follow-up focus group discussion (FGD) among fisherfolks, farmers, and local government officials was also carried out. Responses from the roundtable discussion and FGD were analysed, categorised into themes, and developed into a STEAM research and innovation agenda.

Teacher capacity-building

The second phase of the programme included a workshop for building the capacity of teachers. Specifically, a workshop on Design Thinking Process immersed teachers in several classroom strategies for empathising, problem finding, generating and distilling ideas, prototyping, testing, and communicating to guide STEM students in developing innovations that are responsive to community needs or aligned with local business needs. In addition, the STEAM research and innovation agenda that had been generated from the previous consultation with stakeholders was introduced to the teachers. There was a discussion on how the agenda could be used for the problem-finding stage of the design thinking process.

Innovation-focused curriculum

An SHS curriculum that embeds design thinking was developed to ensure that the innovation programme reached the classrooms. The stages in the design thinking process such as empathising, which entailed understanding home, school, or community needs; problem finding (scoping possible topics based on the STEAM research and innovation agenda); deciding on the focus of the project; and creating and testing prototype solutions ensured innovation in the classroom. Another factor that was observed in the development of the curriculum is its alignment with the existing curricula of the Department of Education for two SHS subjects for the STEM strand, namely: *Inquiries, Investigations, and Immersion* (applied subject) and *Capstone Project* (specialised subject).

Monitoring and evaluation activities

In support of the year-long curriculum implementation, various professional development sessions for teachers, hackathons for students, brownbag sessions on technopreneurship and agricultural innovations, parent education seminars on STEAM, and dialogues with local government leaders on building innovation hubs were conducted.

Pre-tests and post-tests were carried out to assess student learning outcomes such as STEM career readiness (Barnachea & Paderna, 2019) (Appendix), metacognitive awareness (Shraw & Dennison, 1994), grit (Duckworth & Quinn, 2009), growth mindset (Dweck, 1999), and resilience (Smith et al., 2008).

STEAM innovation competition

To spark public interest and support for STEM, as well as plant the seeds for an innovation mindset, the programme culminated in a STEAM Innovation Competition that showcased students' outputs. The original plan was to involve families, young children, local government officials, and other community members in a Community Expo. However, since the COVID-19 pandemic struck, the CISTEM decided to pivot the expo into an innovation competition. For their entries, students were asked to submit explanatory videos of their innovation as well as a digital poster that presents the materials and methods, results and discussion, and conclusion of the research. There were a total of 94 entries spread over the following categories: Agricultural Technology, Life Science, Material Science and Innovation, and Physical Science.

The locale: Sorsogon, Philippines

The STEAM Innovation Programme is implemented in Sorsogon, one of the provinces in the Bicol Region. It is the major producer of pili kernel and a major supplier of fish food in the region. Its agricultural products are among the top priority products of the Department of Trade and Industry (DTI) for agribusiness development. Despite its rich agricultural context, the poverty incidence among families in Sorsogon was reported at 46.2% in 2015 that declined to 26.1% in 2018 (NEDA Region V, 2020). The reason for having more than a quarter of families living in poverty was attributed to its vulnerability to natural hazards such as typhoons, flooding, and volcanic eruption and security threats due to insurgency.

Sorsogon province has 13 public SHS that offer STEM. Twelve of these schools are managed by the Department of Education, while one school is the laboratory school of the local state university.

Initial results of the programme implementation

In January 2019, the Undersecretary of the Department of Education for Curriculum and Instruction and the public school officials of Sorsogon, its superintendents, education programme supervisors, and school principals entered into an agreement to implement the year-long STEAM Innovation Programme in partnership with Unilab Foundation and faculty members from the University of the Philippines-Diliman, College of Education. It was the first of its kind multi-component STEAM innovation programme in the country that involved the community, families, school officials, teachers, and students in various capacity-building opportunities.

Roundtable discussion for defining the STEAM research and innovation agenda

In March 2019, a roundtable discussion (RTD) was conducted involving 70 local community stakeholders of Sorsogon. Participants were school principals, barangay officials, local business owners (e.g., pili oil companies, honeybee farms, pineapple growers, agricultural feed corporations), and representatives from local government agencies such as City/Municipality Planning Offices, Department of Agriculture, Department of Trade and Industry, and Department of Environment and Natural Resources. The RTD aimed to introduce the STEAM Innovation Programme to local community stakeholders and consequently engage them in advancing STEAM education in Sorsogon. During the event, the attendees discussed community and local business needs that were potential areas for STEAM research and innovation in SHS. They also identified forms of support that could be extended to the students, teachers, and schools. Problems that emerged from the RTD included the following: Poor water supply for agricultural activities, limited knowledge on irrigation techniques, traditional farming practices, poor processing of local products, and low interest of the youth in agriculture. An FGD involving some of those present in the RTD was conducted to validate the research and innovation agenda for SHS. The problems that were raised in the RTD and the FGD were considered in the development of the STEAM Research and Innovation Agenda for SHS.

STEAM research and innovation agenda for senior high school

A community needs assessment tool was administered to 486 Grade 12 STEM students. The tool was parallel to the questions that were used in the RTD with local community stakeholders. The tool collected information about common problems that are encountered by students at home, in their school, and in their community. The students were asked to list three common problems at home and in school that need immediate solutions. They were also asked to interview at least two members of their community/*barangay* about problems of the community.

The STEAM Research and Innovation Agenda was culled from the community needs assessment tool that was administered to Grade 12 STEM students and from the outcomes of the RTD and FGD with community stakeholders. Major issues that had been identified in the research and innovation agenda were as follows: Surplus of local produce, innovations without patents, poor weaving skills of the young generation, water irrigation, traditional farming practices, reliance on seasonality of crops, lack of promotion of organic fertilisers, high farming expenses, pest infestations, poor processing of local produce, lack of after-harvest facilities, poor marketing strategies, and conversion of agricultural lands. Problems on disaster risks, connectivity, environmental protection, pollution, sanitation, lack of resources, and safety and security had also been raised. The agenda would be used by teachers and students in identifying problems for their STEM capstone project.

Design thinking workshop for STEM teachers

On May 6–8, 2019, a total of 108 teachers were taught to use the design thinking process for guiding STEM students to generate innovative projects. Teachers were introduced to ideation techniques, invention algorithms, prototyping, and innovation pitches that can be readily implemented in SHS classes. Teachers were also oriented on how to implement the innovation-focused curriculum for their STEM subjects in SHS. Table 7.1 shows Weeks 4–6 and 7–9 of the innovation-focused curriculum.

Table 7.1 Weeks 4–6 and 7–9 of the innovation-focused curriculum

Week	Learning outcomes	Topics	Essential/key questions	Teaching/learning activities	Suggested assessment tools
4–6	• Explain the importance of empathising in the Design Thinking Process • Use empathising strategies for different phases of the Design Thinking Process	Empathise • Aristotle's 7 Elements of a Good Storytelling • What-How-Why Method • Building Empathy with Analogies • 5 Whys • Conducting an Interview with Empathy • Empathy Map	• Why is the Empathise phase important in the Design Thinking Process? • How may each empathising strategy be used for the different phases of the Design Thinking Process	• Activity 1: When Red Balloons Fly • Activity 2: Shrek 2 • Activity 3: Using Analogies • Activity 4: 5 Whys • Activity 5: Conducting an Interview with Empathy	• Activity sheets • Rubric to evaluate group outputs
7–9	• Discuss the importance of problem finding • Use appropriate strategies to identify problems in various contexts	Define (Problem Finding Strategies) • Understanding Contexts • Invitational Stems • Looking at a Problem from a Different Perspective • Thinking of Possibilities and Alternatives • Remixing Words • Bubble Mapping	• Why is it important to learn how to find the right problem for STEM research? • What strategies can be used to identify problems in various contexts?	• Activity 1: My Home, School, and Community • Activity 2: What-If • Activity 3: Be a Bubble Maker	• Activity sheets • Instrument to evaluate group outputs

Table 7.2 Summary of evaluation for the design thinking workshop (*n* = 98)

Item	Mean rating
1 The objectives of the workshop have been attained.	4.88
2 The content of the workshop is relevant to STEM teaching.	4.99
3 The content of the workshop is timely/up-to-date.	4.91
4 The sessions are logically sequenced.	4.91
5 The allotment for each session is reasonable.	4.41
6 The facilitators are knowledgeable on the topics.	4.96
7 The flow of the workshop is systematic.	4.86
8 Workshop materials are well-provided.	4.88
9 Overall evaluation.	4.92
Average Mean Rating	4.86

Additionally, overall evaluation of the workshop (Table 7.2) showed very high satisfaction from the participants with an average rating of 4.92 (with 5 being the highest).

Implementation of the innovation-focused curriculum

In the course of the implementation of the innovation-focused curriculum in schools from June 2019 to February 2020, parent education seminars, continuing professional development sessions for teachers, hackathons for students, online community/curated resources, and dialogues with local government officials regarding an innovation hub were conducted.

It was in March 2020 that the programme was set to culminate in a Community STEAM Expo with the students presenting their research and innovation outputs. There were 94 abstracts submitted for the expo, which were categorised into: (1) Agricultural Technology, (2) Material Science and Innovation, (3) Life Science, and (4) Physical Science. Twelve abstracts per category were selected to advance to the next round. However, the COVID-19 pandemic happened and the Community STEAM Expo had to transition online.

The first communication with DepEd Sorsogon Province and DepEd Sorsogon City regarding the transition of the Community STEAM Expo to an online competition was in May 2020. The 48 qualifying teams were requested to send video presentations and digital posters as entries. Each qualifier submitted a five-minute video presentation to CISTEM, Inc. There were three winners from each category. Non-qualifiers of the contest category automatically competed for the digital poster category, and likewise, three teams were awarded as winners. Winners were announced through CISTEM's Facebook page on July 24, 2020. All the 13 participating STEM schools received an Arduino set of six kits as part of the sustainability plan.

The panel of judges included: Judge 1, an assistant professor of biology education of the University of the Philippines College of Education; Judge 2, a master teacher in a public elementary school; Judge 3, a process innovation specialist in a telecommunication company; Judge 4, a research assistant in a water research and management laboratory of the University of the Philippines; Judge 5, a chemistry research laboratory supervisor; Judge 6, a science research specialist in the Industrial Technology Development Institute of the Department of Science and Technology; and Judge 7, a community lead of the UPSCALE Innovation Hub, University of the Philippines – Diliman.

Following are takeaways of some of the judges:

> I realised with great delight that our students, at a young age, could be very much involved in working out community problems by innovating through design thinking. After the members persevere with the necessary iterations, I hope that their humble projects will become recognised and used across Sorsogon and beyond. – Judge 1
>
> I was truly amazed with the outputs of the students, especially the videos. I can see from their works, all the hard work, creativity, and teamwork they've put in their projects. I also admire the fact that they were able to identify common problems in their community and provide recommendations that are feasible and effective. This only proves that with the right avenue and support from the teachers/school, our students can already provide solutions and make an impact to the community. Hope they will continue what they've started, and that there will be more people supporting these kinds of initiatives to help these students further improve their projects. All the best! – Judge 3
>
> This has been a great learning experience for me not only about the recent researches our young scientists undertake nowadays but also about the vast and rich resources of Sorsogon (e.g. pili, mandarin, balimbing, tawa-tawa, bawang, etc.) that they tried to utilise in their studies.
>
> Also, it is interesting how these young scientists carefully approached their goals through the scientific method and analysed their results with the correct statistical tools. Their creativity in creating their video presentations and digital posters while trying to present their research in a systematic manner is also noteworthy. – Judge 5
>
> It is very fascinating to know that students from Sorsogon practices their creativity using science in order to address various concerns in their community. As a researcher and scientist, I am a firm believer of the great impact of science if applied in terms of various technologies towards the enhancement of human lives. Continue to explore the effects of science and how it can affect our way of living. – Judge 6

Based on the takeaways from the judges, they appreciated the involvement of students in solving community problems by innovating through design thinking. They saw great potential in the programme that provided opportunity for the students to generate feasible and effective solutions that make an impact on the community. They recognised that hard work, perseverance, creativity, and teamwork were evident in the outputs of the students and that the rich local resources were investigated further through the programme.

Future directions

Although the pre-tests were administered at the start of the School Year 2019/2020 in June 2019, administration of post-tests was delayed. This is because classes were suspended in March 2020, and community quarantine was imposed in most parts of the Philippines. Nonetheless, the tests were converted to Google forms and administered online. Collection of responses was recently completed with the data needing to be processed. Test results will then be analysed to determine the effects of the innovation-focused curriculum on students' STEM career readiness, metacognitive awareness, grit, growth mindset, and resilience.

Furthermore, all the components of the programme will be redesigned based on the evaluation results for the sustainability and scalability of the programme. Curriculum guides will be recalibrated and submitted to the Department of Education, Bureau of Curriculum Development for adoption and implementation in other SHS offering STEM. On the other hand, engagement with the STEM teachers will continue through additional professional development activities covering other topics on research and innovation. The experience of the pandemic and subsequent community quarantines brought to light problems on food insecurity and involuntary hunger that are linked to loss of jobs. In response to the said need, the STEAM Innovation Programme will be redesigned to incorporate the development of innovative solutions to combat food insecurity. With the additional component of the programme, the expertise of the Department of Agriculture – Agricultural Training Institute will be tapped, thereby, strengthening inter-sectoral collaborations. Furthermore, since students are staying at home because of the remote learning delivery for the current school year, home-based and community-based innovations will be encouraged.

In terms of the learning resources, the programme will embark on both offline and online materials. Additional materials will have to be developed in consideration of the new delivery modes. Suitable technology platforms will also be explored to support the new setting. Because of restrictions on travelling and gathering, the hackathon, which used to be in-person, will be done through an online platform. The new setup will allow for more mentors to be involved without being constrained by distance, which means that local and international experts can assist the students in their projects. This will consequently grow the innovation ecosystem that the programme wishes to establish with greater participation from government, industry, and other academic institutions.

Conclusion

It was in 2016 that the first batch of SHS students in the Philippines enrolled in their respective schools with more than one million Grade 11 students (Geronimo, 2016; Mateo, 2016). Based on the pre-registration data of the Department of Education as of May 2016, a month before the start of classes for the first batch of SHS students, only 8% of the total number of students had signed up to take up the STEM strand while the majority (40%) followed the technical-vocational track (Banal-Fermoso, 2016). However, in 2018, it was reported that there was a 60% increase in the number of students who developed an interest in STEM throughout the years (Genimiano, 2018). According to studies, students who pursued STEM strands in high schools were significantly more likely to be in a STEM bachelor's degree programme after high school graduation (Means et al., 2018) which is a good manifestation of the improvement in STEM education.

Qualified STEM professionals are needed to remain economically competitive in the global market and to fill contemporary demands such as ensuring sufficient and sustainable energy, efficient healthcare, and well-considered technology development (Bøe et al., 2011). It is estimated that "65% of children entering primary school today will ultimately end up working in completely new job types that don't yet exist" (World Economic Forum, 2016, p. 1). In the Philippines, some of the emerging job roles include software and applications developers and analysts, data analysts and scientists, financial and investment advisers, and database and network professionals (World Economic Forum, 2018). Although Sorsogon is an agricultural province, a considerable number of students' research and innovation projects shows direction toward the said emerging jobs.

A STEAM Innovation Programme is critical in creating a culture of innovation and collaboration. To facilitate its full implementation, important components of such programmes are consultation with stakeholders, teacher capacity-building, an innovation-focused curriculum, learner empowerment, parent education seminars, curated resources, and community expos. A programme redesign is also warranted to ensure its sustainability and scalability. The redesign will include digital conversion of materials and other resources for remote learning delivery, online capacity-building workshops, virtual hackathons, an online community of practice (for teachers and students), and assistance for technology transfer or business development for winning projects.

References

Barnachea, A., & Paderna, E. E. S. (2019). *STEM Career Readiness Scale*. Philippines: University of the Philippines – Diliman, College of Education.

Banal-Fermoso, C. (2016, May 02). *SHS Tracks Offer Career Paths to Students*. Philippine Daily Inquirer: Retrieved from: https://newsinfo.inquirer.net/782629/shs-tracks-offer-career-paths-to-students

Bøe, M. V., Henriksen, E. K., Lyons, T. & Schreiner, C. (2011). Participation in science and technology: Young people's achievement-related choices in late-modern societies. *Studies in Science Education, 47*(1), 37–72.

Commission on Higher Education [CHED] (2019). *CHED Statistics: Higher Education Enrollment by Discipline Group: AY 2009–10 to 2018–19*. Retrieved from: https://ched.gov.ph/2019-higher-education-facts-and-figures/.

Department of Education. (2013). *Suggested Academic Track – Science, Technology, Engineering and Mathematics (STEM) Strand Scheduling of Subjects*. Retrieved from:

https://www.deped.gov.ph/wp-content/uploads/2019/01/Science-Technology-Engineering-and-Mathematics-STEM-Strand.pdf

Department of Education. (2019). *DepEd Welcomes 27.2 M Learners Back to School.* Retrieved from: https://www.deped.gov.ph/2019/06/04/deped-welcomes-27-2-m-learners-back-to-school/

Duckworth, A., & Quinn, P. (2009): Development and validation of the short grit scale (grit–s). *Journal of Personality Assessment, 91*(2), 166–174.

Dweck, C. S. (1999). *Self-Theories: Their Role in Motivation, Personality, and Development.* Philadelphia: Psychology Press.

Feliciano, J., Mandapat, L. C., & Khan, C. (2013). Harnessing the use of open learning exchange to support basic education in science and mathematics in the Philippines. *Online Submission, 3*(6), 407–416.

Genimiano, P. M. (2018, October 24). *More Students Now Prefer Science and Technology Courses.* Philippine News Agency. Retrieved from: https://www.pna.gov.ph/articles/1051947

Geronimo, J. Y. (2016, May 04). *Enrollment for Senior High Begins in Public Schools.* Rappler. Retrieved from: https://www.rappler.com/nation/131717-enrollment-senior-high-school-public-schools

Hallström, J., & Schönborn, K. J. (2019). Models and Modelling for Authentic STEM Education: Reinforcing the Argument. *International Journal of STEM Education.* doi: https://doi.org/10.1186/s40594-019-0178-z10.1186/s40594-019-0178-z.

Honey, M., Pearson, G., & Schweingruber, H. A. (2014). *STEM Integration in K to 12 Education: Status, Prospects, and Agenda for Research.* Washington DC: The National Academies Press.

Imam, O. A., Mastura, M. A., Jamil, H., & Ismail, Z. (2014). Reading comprehension skills and performance in science among high school. *Asia Pacific Journal of Educators and Education, 29*, 81–94.

Lee, M. H., Ching, S. C., & Huang-Yao, H. (2019). STEM Education in Asia Pacific: Challenges and Development. *Asia-Pacific Edu Res, 28*, 1–4.

Mateo, J. (2016, June 18). *Senior High School Enrolment Reaches 1-M Mark.* PhilStar Global. Retrieved from: https://www.philstar.com/headlines/2016/06/18/1594088/senior-high-school-enrolment-reaches-1-m-mark

Means, B., Wang, H., Wei, X., Iwatani, E., & Peters, V. (2018). Broadening participation in STEM college majors: Effects of attending a STEM-focused high school. *AERA Open, 4*(4), 1–17.

National Economic Development Authority Region V (2020). *27 in Every 100 Bicolanos Were Poor in 2018.* Retrieved from: http://nro5.neda.gov.ph/27-in-every-100-bicolanos-were-poor-in-2018/

Official Gazette (2013). *Republic Act No. 10533.* Retrieved from: https://www.official-gazette.gov.ph/k-12/

Sahin, A. (2015). *A Practice-Based Model of Stem Teaching. In a Practice-Based Model of STEM Teaching.* doi: https://doi.org/10.1007/978-94-6300-019-2.

Sansano, A. (2018). *The Philippine Entrepreneurship Ecosystem.* Plenary Presentation at the 2018 Inclusive Innovation Conference, Department of Trade and Industry, Sofitel Plaza, Manila, Philippines, October 2, 2018.

Schraw, G., & Dennison, R. S. (1994). Assessing metacognitive awareness. *Contemporary Educational Psychology, 19*, 460–475.

Smith, B. W., Dalen, J., Wiggins, K., Tooley, E., Christopher, P., & Bernard, J. (2008). The brief resilience scale: Assessing the ability to bounce back. *International Journal of Behavioral Medicine,15*, 194–200.

UNESCO UIS. (2019). Science, Technology and Innovation: *Total R&D Personnel by Sex, Per Million Inhabitants, Per Thousand Labour Force, Per Thousand Total Employment.* Retrieved from: http://data.uis.unesco.org/

Vera Cruz, A. C., Madden, P. E., & Asante, C. K. (2018) Toward cross-cultural curriculum development: An analysis of science education in the Philippines, Ghana, and the United States. In: Roofe C., & Bezzina C. (Eds.). *Intercultural Studies of Curriculum. Intercultural Studies in Education.* Cham: Palgrave Macmillan.

We Are Social. (2018). *Global Digital Report.* Retrieved from: https://wearesocial.com/blog/2018/01/global-digital-report-2018

World Economic Forum. (2016). *Executive Summary the Future of Jobs: Employment, Skills and Workforce Strategy for the Fourth Industrial Revolution.* Retrieved from: http://www3.weforum.org/docs/WEF_FOJ_Executive_Summary_Jobs.pdf

World Economic Forum. (2018). *The Future of Jobs Report.* Retrieved from: http://www3.weforum.org/docs/WEF_Future_of_Jobs_2018.pdf

Appendix: Assessment Tools

Stem Career Readiness Scale

Source: Barnachea, A. and Paderna, E.E.S. (2019). *STEM Career Readiness Scale.* Philippines: University of the Philippines-Diliman, College of Education.

Directions

For each statement, you will have to choose the degree (Strongly Disagree, Disagree, Agree, and Strongly Agree) to which the statement applies to you. Encircle the letter (SD, D, A, SA) corresponding to your answer. Think about each statement independently from other statements. Do not be influenced by your answers to other statements. There is no right or wrong answer. Please respond to all statements.

	Strongly disagree (SD)	Disagree (D)	Agree (A)	Strongly agree (SA)
Science process skills				
1 I am a keen observer.	SD	D	A	SA
2 I can easily compare objects based on their characteristics.	SD	D	A	SA
3 I can measure with great precision and accuracy.	SD	D	A	SA
4 I can conclude based on observations and facts.	SD	D	A	SA
5 I can identify what variables I need to use in my experiments.	SD	D	A	SA
6 I can predict the relationships between variables.	SD	D	A	SA
7 I can discuss results.	SD	D	A	SA
8 I **cannot** formulate a hypothesis.	SD	D	A	SA

(Continued)

	Strongly disagree (SD)	Disagree (D)	Agree (A)	Strongly agree (SA)
9 I can test hypothesis using experiments.	SD	D	A	SA

Mathematical skills

1 I can easily understand mathematical concepts and formula.	SD	D	A	SA
2 I can perform mathematical operations *(e. g. addition, subtraction, multiplication, and division)* with great accuracy.	SD	D	A	SA
3 I am able to use the appropriate formula based on what the problem asks.	SD	D	A	SA
4 I can derive a formula from another formula.	SD	D	A	SA
5 I enjoy solving math problems.	SD	D	A	SA
6 I can explain with confidence how I came up with my answer.	SD	D	A	SA
7 I can defend my solution if I need to.	SD	D	A	SA
8 When I find it hard to solve a math problem, I see it as a challenge and won't give up.	SD	D	A	SA
9 I know how and when I can apply Mathematics in daily-life problems.	SD	D	A	SA

Critical thinking

1 I think of possible results before I take action.	SD	D	A	SA
2 I can make conclusions after observing and gathering facts.	SD	D	A	SA
3 I only make assumptions based on the information given.	SD	D	A	SA
4 I develop my ideas by gathering information.	SD	D	A	SA
5 When facing a problem, I identify options.	SD	D	A	SA
6 I can assess if a statement is logically correct based on information given.				
7 I am able to support my opinions with valid information.	SD	D	A	SA
8 I **cannot** identify which information is relevant to solve a problem.	SD	D	A	SA
9 It is important for me to get information to support my opinions.	SD	D	A	SA
10 I can distinguish if an argument is strong or weak.	SD	D	A	SA

(Continued)

	Strongly disagree (SD)	Disagree (D)	Agree (A)	Strongly agree (SA)
ICT skills				
1 I use valid internet sources to research for information.	SD	D	A	SA
2 I utilise social media to share valid information.	SD	D	A	SA
3 I can select technological resources to accomplish work productively.	SD	D	A	SA
4 I organize my files in my computer.	SD	D	A	SA
5 I name my files in an orderly way so that I can easily search them when I need them.	SD	D	A	SA
6 I can evaluate if an article came from a valid source.	SD	D	A	SA
7 I only share information after validating it.	SD	D	A	SA
8 I think that intellectual property rights should be taken seriously.	SD	D	A	SA
9 I **do not** use technology to threaten other person.	SD	D	A	SA
10 I **do not** practice proper citations when using references.	SD	D	A	SA
Work ethics				
1 I show honesty in my words and actions.	SD	D	A	SA
2 I can be objective in dealing with people and situations.	SD	D	A	SA
3 I believe that I can be trusted with vital information.	SD	D	A	SA
4 If I have issues with someone, I talk to them directly.	SD	D	A	SA
5 Before I make a decision, I consider the pros and cons.	SD	D	A	SA
6 I know when it is appropriate to listen and when to speak.	SD	D	A	SA
7 I do my best in every task that is entrusted to me.	SD	D	A	SA
8 I **cannot** motivate my group mates to achieve a common goal.	SD	D	A	SA
9 I will not do anything that will harm my reputation.	SD	D	A	SA
10 I take full responsibility for my actions.	SD	D	A	SA

Brief resilience scale

Source: Smith, B. W., Dalen, J., Wiggins, K., Tooley, E., Christopher, P., and Bernard, J. (2008). The brief resilience scale: Assessing the ability to bounce back. International Journal of Behavioral Medicine, 15, 194–200.

Metacognitive awareness inventory (MAI)

Source: Schraw, G., & Dennison, R.S. (1994). Assessing metacognitive awareness. *Contemporary Educational Psychology*, 19, 460–475.

Mindset test

Source: Dweck, C. S. (1999). *Self-theories: Their role in motivation, personality, and development.* Philadelphia: Psychology Press.

Short grit scale

Source: Duckworth, A. and Quinn, P. (2009). Development and validation of the short grit scale (Grit–S). Journal of Personality Assessment, 91(2), 166–174.

8 Students' views, attitudes, identity, self-concept, and career decisions
Results from an evaluation study of a STEM program in Singapore

Si Qi Toh, Tang Wee Teo, and Yann Shiou Ong

Introduction

The Prime Minister of Singapore, Mr Lee Hsien Loong, mentioned in his speech at the opening of a local university that Science, Technology, Engineering, and Mathematics (STEM) capabilities are integral for the upgrading of Singapore's economy into a technologically advanced society (Philomin, 2015). To meet the demands of a smart nation, there is a growing imperative to develop inter-disciplinary STEM capabilities in our students to go beyond learning single disciplines in isolation. In recent years, there has been a paradigm shift in how we prepare the future generation for STEM occupations to meet the demands of a restructured economy. This includes a greater focus on authentic learning experiences that involve using hands-on learning and technological tools to solve problems across multiple disciplines. The manifestation of a strong narrative for STEM is the emergence of STEM education programmes to develop students who are able to apply the knowledge and skills from across the STEM disciplines through authentic contexts. Some topics include robotics, healthcare, forensic science, and water technology.

While the idea to offer STEM curriculum is laudable, implementation issues ensue. Earlier studies (see e.g., Dong et al., 2020; Johnson, 2012) have reported on the challenges of STEM implementation in schools due to the lack of curriculum time, relevant content knowledge of teachers, and resources for implementation. In order to promote large-scale implementation of STEM in Singapore schools, the Ministry of Education has offered STEM as an Applied Learning Programme (ALP) for students to gain authentic learning experiences and opportunities to integrate knowledge and skills from across the STEM disciplines. Since 2014, more than 50 Singapore secondary schools (equivalent to Grades 7–10) have embarked on the STEM ALP that provides access to financial and intellectual capital for implementation. However, STEM ALP is a programme rolled out only to MOE schools (i.e., publicly funded, non-independent, and non-autonomous). Some private and independent schools in Singapore have also designed their own STEM programmes for their students.

DOI: 10.4324/9781003099888-8

At the time of this study, there were no reports of an evaluation study of the STEM programmes in Singapore. This research seeks to measure the effectiveness of the STEM programmes offered in Singapore schools. Based on the literature in STEM education research, STEM learning outcomes may be broadly categorised as students' attitudes towards STEM lessons, views about STEM, student's self-concept while learning STEM, construction of STEM identities, and future aspiration in STEM-related fields. These categories form the five constructs in the survey instrument that was developed to measure the outcomes of the STEM programmes implemented in Singapore schools. The findings offered useful insights for fine-tuning such programmes in schools.

STEM programmes in Singapore

Many schools in Singapore have decided to introduce their students to STEM through STEM exposure trips and annual STEM weeks. Beyond exposure, some schools have also taken the additional step to implement STEM applied programmes in their curriculum. To cite some examples, independent schools such as Methodist Girls' School (MGS), Raffles Institution (RI), Raffles Girls' School (RGS), and Hwa Chong Institution (HCI) have implemented their own STEM applied programme focused on developing the STEM competencies (e.g., problem-solving skills and critical thinking skills) of their students. In some of these independent schools, the STEM programme may only be intended for high-ability students. For example, HCI has its own science and mathematics talent development programme (SMTP) that aims to integrate science and mathematics disciplines to deepen the learning of a chosen group of students. Similarly, RI and RGS offer the Raffles Academy programme, in which selected students would be able to take up interdisciplinary research or coursework modules. On the other hand, MGS offers a STEM enrichment programme compulsory for all lower secondary students with topics such as coding and robotics. Having recognised the importance of having STEM integrated into their curriculum, some government schools have also decided to implement compulsory STEM applied programmes for lower secondary school students. Due to the novelty of STEM, schools often develop a partnership with Science Centre Singapore for expertise in crafting STEM curriculum and providing STEM-trained instructors to facilitate the programme. The Science Centre Singapore has a subsidiary known as STEM Inc that oversees the development of STEM programmes in secondary schools. STEM Inc has developed STEM programmes in 12 themes (robotics, alternative energy, applied health sciences, etc.) with over 40 projects that schools can choose to embark on (Science Centre, n.d.). For our research, we will examine STEM applied programmes implemented in Singapore public schools.

Given the diversity of the STEM programmes implemented in schools, the STEM education programme discussed in this paper is referred to as STEM-X to collectively refer to the type of STEM programmes that offer applied learning experiences for students.

STEM-X

STEM-X takes on different forms in Singapore schools. They differ in the themes, foci, outcomes, and approach. In one case, a school selected "food science and technology" as a theme and designed the programme for students to develop empathy through identifying the needs of others, develop prototypes that solve a problem encountered by a group of needy people, and think of ways to scale up the prototype to benefit more people. Another school combined their STEM programme with community service work in planning healthy and professional meals for elderly people. This school aimed to develop students' soft skills such as collaborative skills, interview and surveying skills, presentation skills, and empathy. They also want students to acquire technological and simple designing skills. Another school has also challenged students to apply their knowledge and skills related to smart technologies to design solutions related to traffic and mobility of the elderly and less-abled.

While the themes, goals, and implementation of STEM-X may differ across schools, they are similar in a few ways. Firstly, the students are engaged in the learning and exploration-based upon a real-world context or problem. Such problems are engaging for students because they are either problems students have encountered in their personal lives or problems that are novel to students, such as constructing a robot and making it move or complete a task. Secondly, the activities require students to identify solutions for the benefit of a group of people and not just for themselves. In order for students to adequately address the problem, students have to first understand the needs of the people who are the beneficiaries of their STEM solutions. Thirdly, students engage in collaborative work in the activities as the problem tends to be complex, multi-faceted, and the solutions are not obvious. In essence, the STEM-X approach is somewhat similar to problem-based learning (Hmelo-Silver, 2004) approach except that the content leverages STEM-related concepts.

Theoretical framework

STEM capital

Drawing on Bourdieu's theory on the forms of capital (1986), Archer et al. (2015) extended the concept to science by proposing the idea of science capital to predict student's engagement and aspiration towards science. Adding on to Archer's definition of science capital, Teo and Goh (2019) included students' mental schema as an additional aspect to broaden the definition of science capital as predictors of science participation. Mental schemas would take into consideration students' views towards their science lessons, their self-concept through learning STEM, which ultimately culminates in the development of their self-identity (a component of science capital concerned with the constructions of self-views about who one is and in relation to others) and aspiration towards STEM.

Measuring STEM capital

In a study to examine the psychometric properties of the factors affecting students' career aspiration in science, Jones et al. (2020) found four predictors of science capital – Science Expectancy Value, Future Science Task Value, Science Experience, and Family Science Achievement. Science Expectancy Value measures students' self-concept and self-efficacy, as well as their perception of the benefits of pursuing a career in science. Future Science Task Value assesses students' views on the transferability and applicability of scientific knowledge. Next, Science Experience measures students' attitudes towards their own science lessons and the availability of science resources. Lastly, Family Science Achievement measures the importance the family places on science.

Drawing on the previous research (see e.g., Archer et al., 2015; Teo & Goh, 2019) on science identity and science capital, we propose five constructs related to STEM capital to serve as measures of the effectiveness of the schools' STEM-X in this research.

The five research constructs examined (not in order of importance) are:

1 Students' views about STEM lessons
    This construct measures students' general interest during STEM lessons, views about the difficulty of the lesson and the impact of these lessons on the students' aspirations towards a career in STEM. Renninger (2000) suggested that interest can lead to a long-term, prolonged engagement with a subject. However, researchers have been clear to distinguish between two forms of interest – individual and situational interest (Hidi, 2001). Students' views about STEM lessons constitute situational interest as positive feelings are often associated with the STEM lesson rather than the subject itself. However, situational interest maintained over a long period of time can potentially result in individual interest which is defined as an enduring predisposition towards a certain domain (Hidi & Renninger, 2006). In the context of STEM-X, this construct also measures if the interest evoked during lessons translates to sustained interest in STEM beyond the programme itself.
2 Attitudes towards STEM
    Students' views towards their STEM lessons would also influence their attitudes and feelings towards STEM. Wigfield and Eccles (2000) demonstrated how the distinct beliefs students hold of the value of certain occupations can in turn affect their proclivity towards certain occupations. Following Jones et al. (2020)'s Future Science Task Value, students' positive perception towards the utility and applicability of STEM would then contribute to their inclination towards STEM fields. This includes students' views on the necessity of STEM in a future occupation as well as the importance of STEM in our society.
3 Self-concept in learning STEM
    Pivotal to students' continual engagement in STEM is the development of a healthy self-concept in relation to STEM. Self-concept in science refers

to how a student perceives his or her own scientific abilities based on their experiences and the extent to which they are being recognised by others for their scientific abilities (Shavelson et al., 1976). According to Carlone and Johnson (2007), science engagement and participation is heavily dependent on self-concept as the extent to which a student feels recognised and validated for their science ability would affect their self-efficacy which is referred to as students' belief in their ability to succeed in tasks and courses (Bandura, 1982). As researchers posit, self-efficacy would then directly influence students' future engagement and accomplishment in STEM (Bandura, 1997). Hence, a high self-efficacy would also indicate a higher likelihood of participating in more STEM tasks and eventually pursuing a career in STEM (Lent et al., 1984).

4 Construction of STEM identities

Central to the discussion on engagement and persistence in STEM is the construction of STEM identity. Similar to science identity, STEM identity indicates one's perception of their belonging in the STEM community (Carlone & Johnson, 2007). STEM identity is largely influenced by recognition and individual interest (Hazari et al., 2010).

Beyond situational interest developed through having positive attitudes towards STEM lessons, the solidification of one's identity in STEM would contribute to prolonged engagement in STEM, aspiration towards STEM fields and alignment through concrete actions taken by students towards furthering their interest in STEM beyond STEM-X (Wenger, 1998), which further solidifies their perception of their sense of belonging in the STEM community.

Hence STEM self-concept, views towards STEM lessons and attitudes towards STEM all contribute to the development of STEM identity. Identity can then account for persistence and achievement in studying STEM (Dou et al., 2019).

5 Career decisions in STEM

As STEM occupations are expected to contribute significantly to Singapore's economy, it is important to measure the extent to which STEM-X is effective in encouraging students to aspire towards STEM occupations. Beyond attracting students to join STEM careers, students' responses would also give us key insights to retaining individuals in STEM, given that many individuals stop pursuing STEM early into their undergraduate studies (Wilson et al., 2012).

According to Eccles and Wigfield's expectancy-value theory (2002), an individual's engagement in STEM tasks and eventual career choice can be explained by four components; attainment value, intrinsic value, utility value, and cost. Attainment value reflects how well one perceives a completion of a task would confirm their identity and is highly related to self-efficacy and STEM identity. Intrinsic value refers to the enjoyment one receives from participation in the task, similar to Deci and Ryan's (1985) intrinsic motivation stemming from individual interest. Utility value is determined by one's perception of the benefits of STEM to the individual and to society. In contrast to the other components, cost refers to the negative aspects that the individual considers such as the amount of effort required to succeed, societal expectations. Hence, this construct measures whether views about STEM lessons, attitudes towards STEM, self-concept, and identity in STEM have culminated in students having an increased inclination towards STEM careers.

Although expectation-value theory explains the effect of an individual's mental schema on their career choices, yet one's aspiration towards STEM fields can also be highly influenced by cultural capital such as family attitudes, views, and support for the sciences. Researchers have noted that perceived family values and attitudes towards science influence students' self-concept and individual interest eventually influencing future career decisions (Jones et al., 2020). Hence, it is necessary to also measure students' family's perception of STEM occupations.

Methods

We identified 60 schools that implemented STEM-X for a duration of up to about six months. All of the schools identified had in-house programmes, with some schools partnering with external STEM providers to run the lessons. Email invitations were sent to the 60 schools. Out of the 60 schools, 13 schools expressed interest in research participation, giving us a response rate of 21.7%. Once schools consented to participate in the study, they were requested to assist in the recruitment of 50 students who had participated in STEM-X. Email invitations were sent to students who had given assent and parental consent to participate in the study to complete an online survey. Participation was voluntary and since the survey was administered at the end of the school year, it was difficult to follow up with the students to ensure that they had completed the survey. Eventually, a total of 151 Secondary 1 or 2 (equivalent to Grades 7–8; aged 13–14) students from 13 schools participated in this study.

All 13 schools implemented STEM-X programmes for Secondary 1 and/or 2 students. However, for our study, our target group surveyed were the lower secondary students who went through the cohort-wide programme that was part of the school's formal curriculum. Most schools chose to focus on one specific applied STEM topic such as robotics, aerospace engineering, coding, health science, and food science. Many also indicated their partnership with external STEM providers.

As gathered from information provided on these schools' websites about STEM-X, the schools stressed the importance of authentic learning, boasting programmes with hands-on activities and learning journeys with industry partners. Beyond exposure to STEM and development of STEM capabilities, students are also required to collaborate with one another to solve problems.

Survey items were adapted and expanded from a survey created by Teo and Goh (2019) for the measurement of science capital according to the above-mentioned research constructs. For each item, students were asked to respond according to a five-point Likert-scale rating (Strongly disagree = 1, Disagree = 2, Neutral = 3, Agree = 4, Strongly Agree = 5).

Instrumentation and analysis

Through WINSTEPS, Rasch analysis was employed to analyse the data collected. Rasch analysis involves inferential statistics which allows us to order the respondents according to their agreeability and items according to their difficulty. Hence, person agreeability is taken into consideration when ranking the item difficulty and person and item misfits can also be eliminated in the process. A person-item Wright map was generated for each construct and used to make inferences. Negatively worded items (marked with an asterisk) were flipped prior to analysis.

Findings and discussions

Construct A: Attitude towards STEM lessons

Analysis showed that the person mean ("M" on the left side) is 1.2 logits above the item mean ("M" on the right side of the Wright map, Figure 8.1), this indicates that students generally had positive views of their STEM lessons.

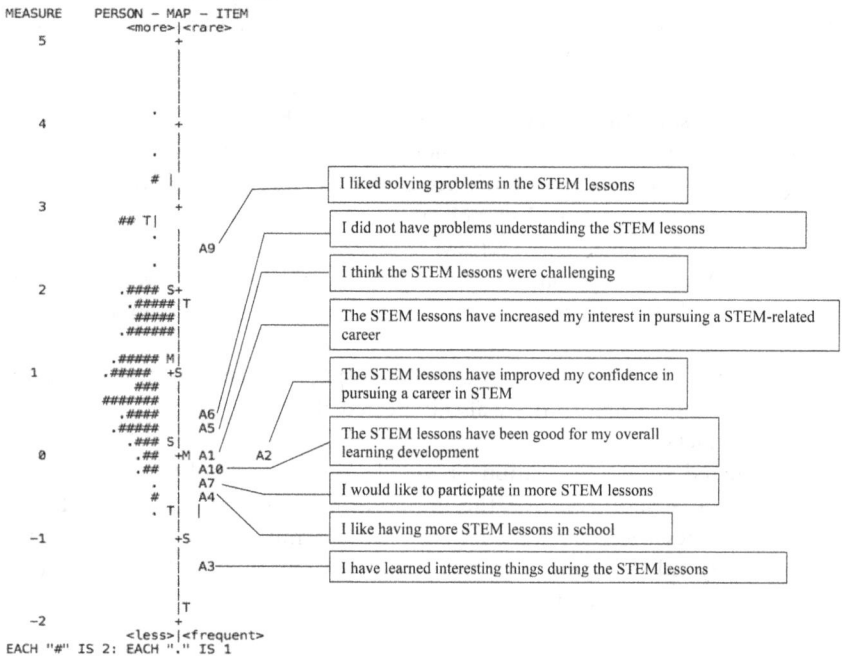

Figure 8.1 Wright map of students' attitude towards STEM lessons.

In reference to the item mean, A5, A6, and A9 fell above the item mean, which indicate that students found these items relatively harder to agree with. In comparison, items A1, A2 were on the mean while A10, A7, A4, A3 which had fallen below the item mean were relatively easier to agree with.

Although students expressed having difficulty understanding the content of the lessons (A6), they did not perceive the lessons to be too challenging (A5). This suggested that the students were well-supported in the lessons. Hence, this could have contributed to students being engaged during STEM lessons (A3) as well as encouraged the development of students' STEM skills. Furthermore, many students also indicated that their STEM lessons piqued their interest and confidence in pursuing a career in STEM (A1, A2). Overall, the students had favourable views towards their STEM lessons as evident by majority of the students expressing their desire for more STEM lessons in their school curriculum (A7, A3). This is important in the development of situational interest in STEM, which preludes individual interest (Hidi, 2001).

However, about 90% of the students had difficulty agreeing to item A9 about solving problems (refer to Figure 8.1). Despite thinking positively of STEM lessons, the students were generally averse to solving problems during STEM lessons. Although problem-based learning (PBL) may be a promising approach for STEM-X, implementational challenges include teacher and student readiness as well as instructional design (Tan, 2004). Students may prefer more traditional pedagogical approaches due to their continual exposure to other lessons. Hence, as compared to the more predictable format of traditional pedagogical approaches, the format of PBL may be confusing to students if not conveyed properly and continuously throughout the course (Tan, 2004). Along with student readiness, if teachers are not well trained in PBL, they might face challenges in scaffolding and facilitating PBL. This could in turn contribute to students' confusion which may also affect students' self-efficacy (Tan, 2004) and the development of individual interest in STEM.

Construct B: Views on STEM

Referring to Figure 8.2, the person-mean is 1.4 logits higher than the item mean, this indicates that majority of students agreed on the importance of STEM.

Items B4, B7, B9, B10, B1, B3, and B2 fell below the item mean, indicating that students found these items easier to agree with, as compared to items B8*, B5, B6. In general, students had positive feelings towards STEM (B1) as they recognised the impact of learning STEM on their personal abilities and skills (B10, B9). Also, students believed in the applicability and transferability of STEM knowledge in solving most real-life issues (B2, B3, B7, B5), Although, at times students struggled with understanding STEM (B8), most students also disagreed with the idea that STEM knowledge and skills are only accessible to highly trained individuals (B6).

The positive views on the transferability of STEM education is encouraging as both Archer et al. (2015) and Jones et al. (2020) reported that the transferability

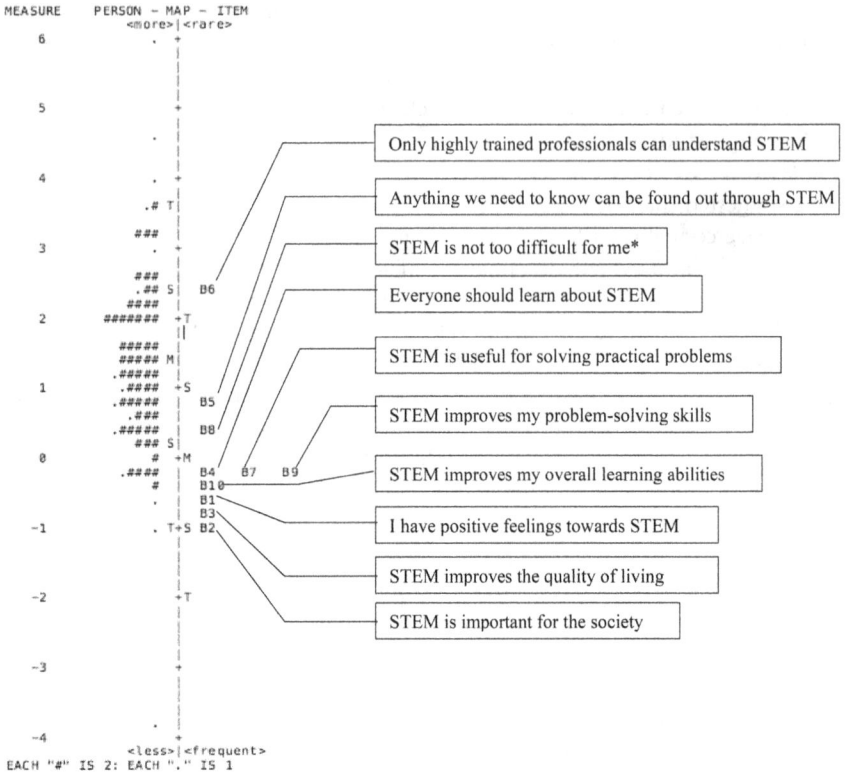

Figure 8.2 Wright map of students' views on STEM.

of science is one of the strongest predictors of future aspiration and their identity as science students. The students' favourable view on STEM seems to be an indication of the success in STEM-X in this aspect. Perhaps, through STEM-X students were able to gain a better understanding of the importance of STEM.

Construct C: Self-concept in learning STEM

Referring to Figure 8.3, the person mean is 0.6 logits higher than the item mean. This indicates that generally, students agreed they were confident of their abilities during STEM lessons.

Items C7, C9, C10*, C4, C1 which fall below the item mean were easier to agree with as compared to C2*, C5*, C3, C6, C8. Looking at the ranking of item difficulty, it is interesting to note that although 90% of students agreed that they were able to do well during STEM lessons (C4) and that learning STEM is not too difficult for them (C10*), items related to students' comparison of their own STEM abilities to other students' abilities or to their skills in other domains (C8, C6, C3) were harder to agree with.

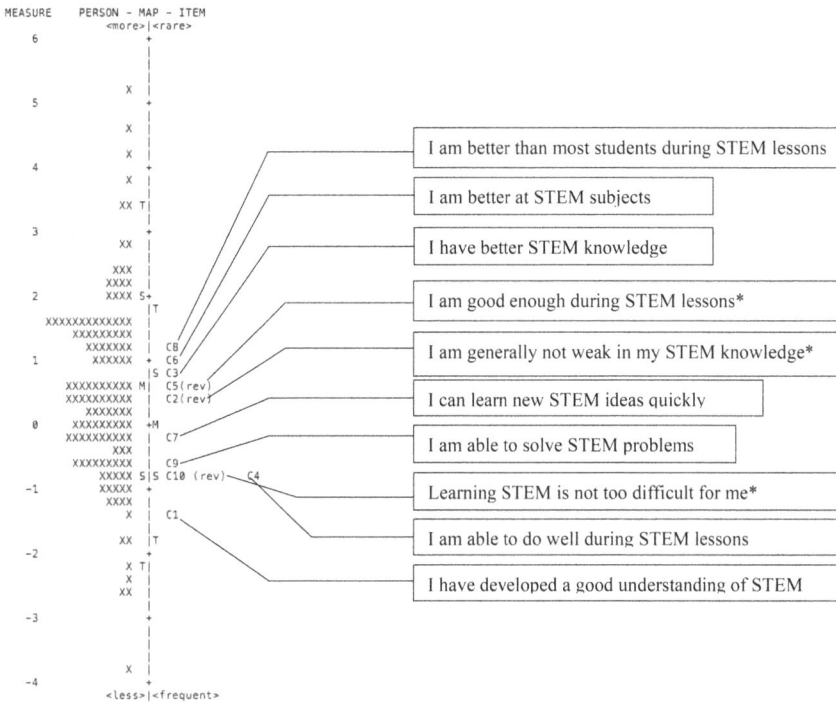

Figure 8.3 Wright map of self-concept in learning STEM.

Although students agreed that they are able to do well during STEM lessons (C4), they believed that they were not good enough in comparison to their peers (C5*). Since their self-concept is formed by comparisons of their own ability to other students during STEM-X lessons as well as between different subjects (Cooper et al. 2018), despite expressing confidence in their ability to learn and solve STEM problems (C7, C10*) students' assessment that they are less capable than their peers suggest a lack of formation of positive STEM self-concept through the lessons. Self-concept would affect one's self-efficacy in STEM, which is defined as students' perception of how well they will perform in STEM (Bandura, 1982). Students with low self-efficacy in STEM subjects are then less likely to continue in STEM. Consequently, in the future, this also influences their academic and career choices (Lent et al., 1984), as students with low self-efficacy would be less likely to continue in the field.

Construct D: Construction of STEM identities

Referring to Figure 8.4, person mean is 0.6 logits higher than item mean, indicating that students agreed that STEM-X positively affected their perception of their belonging in the STEM community. Items that fell below the item mean were easier to agree with, were D7*, D2*, D6, D3*, D4*, D9 while D1, D5,

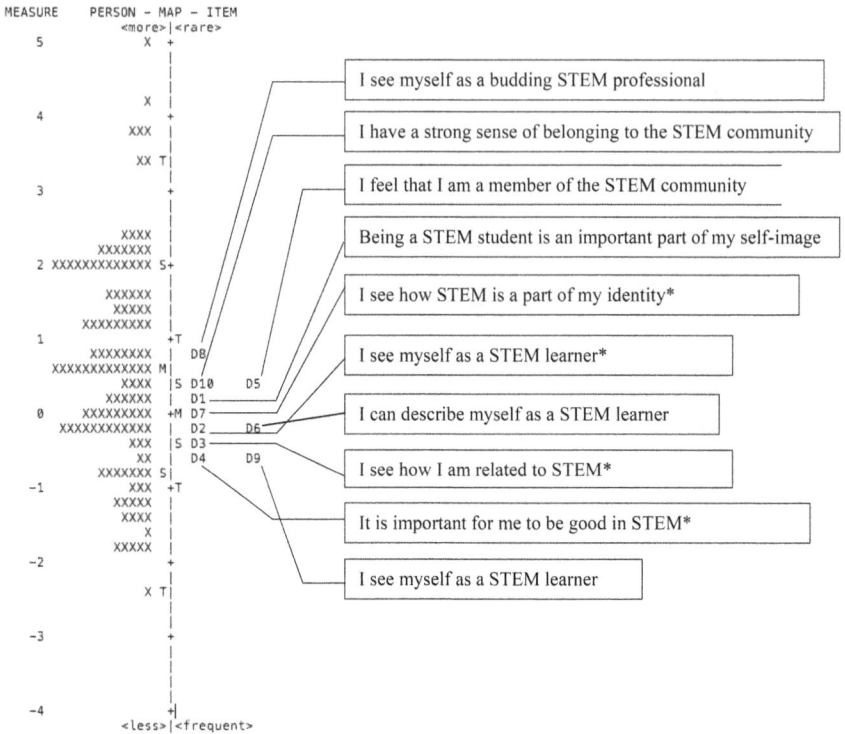

Figure 8.4 Wright map of construction of STEM identities.

D10, D8 fell above the item mean with D8 being the hardest item to agree within this construct. On opposite ends of the spectrum were "I see myself as a STEM learner" (D9) which is the easiest for the students to agree with and "I see myself as a budding STEM professional" (D8) which is the hardest to agree with. Along with this, items relating to students' perception of their sense of belonging in the STEM community were also harder items to agree with (D10, D5).

The results seem to indicate that students viewed themselves as consumers or learners of STEM (D2*, D6, D9). Although students recognised the importance of learning and excelling in STEM (D4*), yet they do not see themselves as producers and contributors of STEM. Wenger (1998) posits three levels of belonging that contribute to the formation of identities – *engagement, imagination, and alignment. Engagement* refers to participation and situational interest in STEM, which occurs through programmes such as STEM-X. Moving forward, *imagination* would refer to students envisioning their continual participation due to individual interest. Lastly, *alignment* would refer to students taking concrete actions to align themselves with what they view members of the community would do (i.e., embarking on personal STEM research beyond school curriculum). Although students agreed that STEM formed a part of their identity (D7*), students' responses seem to indicate that a level of identity that does not proceed beyond *engagement*. Students' interest in STEM remains a situational interest as it seems to be demonstrated only during their participation in STEM-X.

Construct E: Career decisions in STEM-related fields

According to Figure 8.5, the person mean is 0.4 logits higher than the item mean, which indicates that students were generally divided on whether to pursue STEM careers. Items that fell below the item mean were E1, E10, E2, E3, E6, E8*, which items that fell above the mean were E7, E4, E5*, E9.

Most students recognised the benefits of a career in STEM (E8*) and many expressed interests in a career in STEM (E1, E2). Although students did not perceive STEM occupations to be more prestigious than any other careers (E7), they recognised the competitiveness (E10) and high financial rewards of STEM occupations (E6). According to Lent et al. (2003), outcome expectations are a component of social cognitive framework that influence career choices.

Outcome expectations include students' perception of financial benefits, social approval, and personal satisfaction of choosing a career in STEM. From the survey results, we gathered that financial benefits and social approval may be the main driving factors of students having aspirations in STEM careers (E9). Despite this, many students do not foresee having personal satisfaction in STEM careers (E9). Again, the lack of individual interest could be affected by other

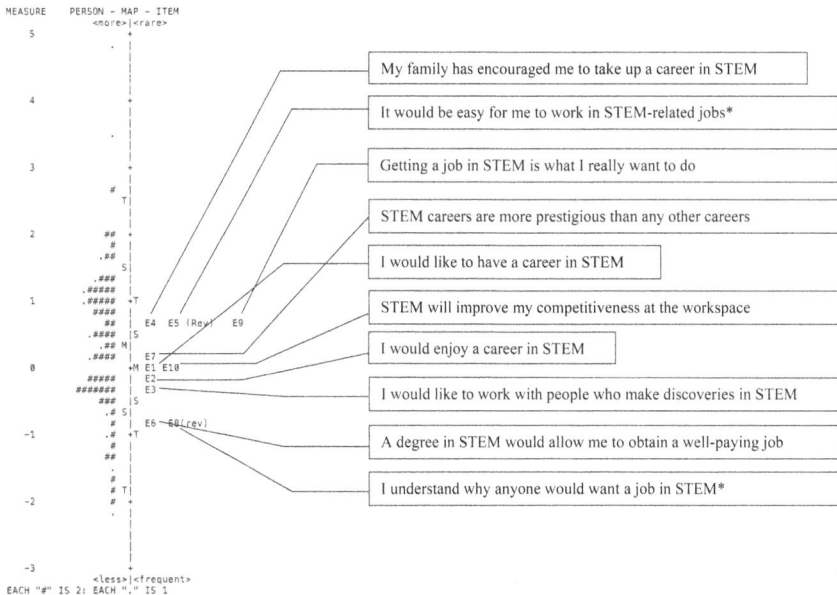

Figure 8.5 Wright map of career decisions in STEM-related fields.

factors such as self-concept, execution of pedagogical practices, or views about STEM. Also, the perception that STEM occupations are too difficult (E5*) could also be a result of low self-efficacy as reported in the analysis of students' self-concept through learning STEM.

Additionally, according to Archer et al. (2015) and Jones et al. (2020), family influence is a strong predictor of whether a student is likely to pursue careers in STEM. This could be due to parents being unfamiliar with STEM occupations due to the novelty of STEM. Students with families which support a career in STEM would be encouraged to continually participate in STEM endeavours and hence would have a greater likelihood of having aspirations towards STEM occupations (Jones et al. 2020). Hence, lack of parental encouragement may be a hindrance in students' aspirations towards STEM occupations. However, although it is a salient factor that affects the formation of STEM identities and science aspiration, parental influence is largely out of the realm of control of STEM-X.

Implications for future research and limitations

Limitations of Rasch analysis

Although only 151 students were surveyed, since there were only 10 items per construct, only 100 participants were needed for a sufficient sample size. Item separation reliability of all constructs exceeded 0.90 which indicates that each construct has items with a wide range of difficulty. On the other hand, the person separation reliability for each construct fell between 0.60 and 0.90 which indicates a heterogeneous sample of students surveyed (refer to appendix).

Rasch model assumes that all items were perceived in the same order of difficulty to all participants. However, due to differences in perspectives and interpretations of the items, item difficulty would vary among different individuals. As the Rasch analysis is made based on the ranking of person agreeability and item difficulty, varying interpretations of item difficulty may result in a decreased reliability of the data. Hence, it is important to look at the item and person fit to eliminate any misfitting data.

Item A8 (I like STEM lessons) had an infit of 2.17 and outfit of 3.06, which extended beyond the conventionally acceptable range for a t distribution of 3.00 (refer to Table 8.1.). Hence, we removed item A8 and the data was re-analysed using Rasch and the infit and outfit of the remaining items are reassesed (refer to Table 8.2.) to be less than 3. For future studies, we will recruit a large sample size of participants to check if this item should be included in the construct.

Recommendations and implications

Through these findings, we were able to identify some issues that hinder students' engagement and aspiration towards STEM. This has implications for STEM task design and implementation. When relying on the predictors of

Table 8.1 Item statistics misfit order for Construct A (Item A8 not removed)

Input: 143 Person 10 Item Reported: 143 Person 10 Item 5 Cats WINSTEPS 3.80.1

Person: Real sep.: 1.34 Rel.: .64 ... Item: Real Sep.: 6.64 Rel.: .98

Item statistics: Misfit order

Entry Number	Total Score	Total Count	Measure	Model S.E.	Infit		Outfit		PT Measure-A		Exact EXP.	Match		Item
					MNSQ	ZSTD	MNSQ	ZSTD	CORR.			OBS%	EXP%	
8	325	143	1.97	.10	2.17	8.2	3.06	9.9	A	.13	.58	35.0	40.9	A8
5	502	143	.34	.10	1.60	4.4	1.81	5.7	B	.09	.54	39.2	45.9	A5
6	495	143	.41	.10	1.09	.8	1.10	.9	C	.46	.54	40.6	44.9	A6
3	620	143	-1.16	.13	.92	-.6	.84	-1.2	D	.56	.43	57.3	55.2	A3
10	585	143	-.63	.12	.72	-2.4	.79	-1.8	E	.60	.48	60.1	52.3	A10
7	559	143	-.29	.11	.77	-1.9	.76	-2.1	e	.74	.50	57.3	50.1	A7
4	572	143	-.46	.11	.68	-2.9	.65	-3.2	d	.77	.49	56.6	50.7	A4
9	547	143	-.15	.11	.65	-3.3	.66	-3.2	c	.60	.51	58.0	49.2	A9
1	540	143	-.07	.11	.56	-4.3	.57	-4.2	b	.70	.52	63.6	48.3	A1
2	529	143	.05	.10	.54	-4.6	.55	-4.5	a	.74	.52	65.0	48.1	A2
Mean	527.4	143.0	.00	.11	.97	-.7	1.08	-.4				53.3	48.6	
S.D.	76.2	.0	.79	.01	.50	3.9	.75	4.4				10.3	3.8	

Table 8.2 Item statistics misfit order for Construct A (Item A8 removed)

Input: 143 Person 10 Item Reported: 143 Person 10 Item 5 Cats WINSTEPS 3.80.1

Person: Real sep.: 1.34 Rel.: .64 ... Item: Real sep.: 6.64 Rel.: .98

Item statistics: Misfit order

Entry Number	Total Score	Total Count	Measure	ModelS.E.	Infit		Outfit		PT Measure-A		Exact	Match		Item
					MNSQ	ZSTD	MNSQ	ZSTD	CORR.		EXP.	OBS%	EXP%	
8	299	136	2.42	.11	2.00	6.9	2.10	7.5	A	.25	.58	33.1	44.7	A9
5	480	136	.37	.11	1.52	3.8	1.64	4.6	B	.14	.53	40.4	48.9	A5
6	474	136	.45	.11	1.30	2.3	1.30	2.3	C	.37	.53	41.2	48.0	A6
3	593	136	-1.35	.14	.95	-.4	.90	-.7	D	.49	.43	60.3	58.5	A3
7	538	136	-.41	.12	.81	-1.6	.78	-1.9	E	.71	.49	59.6	52.7	A7
4	548	136	-.56	.12	.70	-2.6	.69	-2.0	e	.74	.48	58.1	53.0	A4
9	520	136	-.15	.12	.63	-3.4	.66	-3.1	d	.59	.50	58.8	51.8	A10
1	515	136	-.08	.12	.62	-3.5	.63	-3.5	c	.65	.51	64.7	51.1	A1
2	508	136	.01	.12	.53	-4.7	.52	-4.8	b	.73	.51	66.2	51.0	A2
Mean	503.1	136.0	.00	.12	.98	-.6	1.00	-.4				54.6	51.4	
S.D.	75.9	.0	.95	.01	.46	3.6	.49	3.7				11.1	3.5	

STEM Capital as a tool for evaluating the success of STEM-X, cultural capital such as the family support for the STEM may be beyond the realm of control of curriculum designers and implementers. Although students were found to be more aware of the application of STEM as well as its positive impacts on society as a result of STEM-X, students' low self-concept and disdain towards problem-solving are worrying factors. As both factors are not mutually exclusive and may reinforce one another (Tan, 2004), moving forward, STEM-X curriculum designers and implementers should focus on developing student's self-concept and help students grow more accustomed to problem-based learning.

Given that the purpose of an applied STEM programme is to equip students with the skills and knowledge to solve complex, real-life problems, there is a need to better cultivate problem-solving competencies. As one of the primary reasons for students' disinclination towards problem solving is poor task design (Tan, 2004), this can be improved through adopting the S-T-E-M quartet instructional framework (Tan et al., 2019) which improves on the Problem-Based Learning (PBL) framework by defining the authenticity of a problem by its relevance to the STEM disciplines as well as its practical value.

According to the S-T-E-M quartet, authentic problems should be persistent, complex, and extended. Persistent problems refer to problems with explanations that can be applied in a wide variety of contexts. Complexity refers to the necessity of drawing knowledge from two or more STEM disciplines to solve the problem. Extended problems refer to the need for students to have prolonged engagement with the problem (recommended three to five hours to propose a solution). While emphasising the horizontal connections between the STEM disciplines, the STEM curriculum should have one lead discipline. For example, in an applied STEM programme on health sciences, science may be the lead discipline with knowledge on human biology being the necessary foundation to which disciplines like technology and engineering are needed for the creation of an authentic problem. Through the use of this framework, curriculum designers and implementers would be able to effectively assess and create authentic STEM tasks.

Beyond task design, to support student's development of self-efficacy in STEM, curriculum implementers should focus on providing mastery experiences by tailoring each task to individual students' developing abilities while providing the necessary scaffolding. Beyond mastery experiences, it is also important for teachers to leverage social persuasion by encouraging giving genuine and realistic encouragement, whilst helping students navigate anxieties that would arise when engaging in problem solving (Britner & Pajares, 2005)

Our research presents the general feelings and attitudes of students from 13 schools towards STEM-X. Although this allows us to gauge the general sentiments of students, due to the difference in interpretation and implementation of STEM among the different schools, each school would have their own distinct challenges. Future researchers should consider evaluating the pedagogical practices of the STEM-X in each school to determine how to best facilitate learning as well as to provide more targeted feedback to individual schools.

Acknowledgements

Approval to conduct this study was granted by the Nanyang Technological University (Singapore) Institutional Review Board (IRB-2019-03-033) and Ministry of Education (EDUN N32-07-005).

References

Archer, L., Dawson, E., DeWitt, J., Seakins, A., & Wong, B. (2015). 'Science capital': A conceptual, methodological, and empirical argument for extending Bourdieusian notions of capital beyond the arts. *Journal of Research in Science Teaching, 52*(7), 922–948. doi: 10.1002/tea.212270033-z.

Bandura, A. (1982). Self-efficacy mechanism in human agency. *American Psychologist, 37*(2), 122–147. doi: 10.1037/0003-066X.37.2.122.

Bandura, A. (1997). *Self-Efficacy: The Exercise of Control.* WH Freeman/Times Books/ Henry Holt & Co. Westport, CT: Greenwood.

Bourdieu, P. (1986). Forms of capital. In Richardson J. (Ed.), *Handbook of Theory and Research for the Sociology of Education* (pp. 241–258). New York: Greenwood.

Britner, S., & Pajares, F. (2005) Sources of science self-efficacy beliefs of middle school students. *Journal of Research in Science Teaching, 43*(5), 484–499.

Carlone, H. B., & Johnson, A. (2007). Understanding the science experiences of successful women of color: Science identity as an analytic lens. *Journal of Research in Science Teaching, 44*(8), 1187–1218.

Cooper, K. M., Krieg, A., & Brownell, S. E. (2018). Who perceives they are smarter? Exploring the influence of student characteristics on student academic self-concept in physiology. *Advances in Physiology Education, 42*(2), 200–208.

Deci, E. L., & Ryan, R. M. (1985). The general causality orientations scale: Self-determination in personality. *Journal of Research in Personality, 19*(2), 109–134.

Dong, Y., Wang, J., Yang, Y. et al. (2020). Understanding intrinsic challenges to STEM instructional practices for Chinese teachers based on their beliefs and knowledge base. *International Journal of STEM Education, 7,* 47. doi: https://doi.org/10.1186/ s40594-020-00245-0.

Dou, R., Hazari, Z., Dabney, K., Sonnert, G., & Sadler, P. (2019) Early informal STEM experiences and STEM identity: The importance of talking science. *Science Education, 103,* 623–637.

Hazari, Z., Sonnert, G., Sadler, P. M., & Shanahan, M. C. (2010). Connecting high school physics experiences, outcome expectations, physics identity, and physics career choice: A gender study. *Journal of Research in Science Teaching, 47*(8), 978–1003.

Hidi, S. (2001). Interest, reading, and learning: Theoretical and practical considerations. *Educational Psychology Review, 13*(3), 191–209.

Hidi, S., & Renninger, K. A. (2006). The four-phase model of interest development. *Educational Psychologist, 41*(2), 111–127.

Hmelo-Silver, C. E. (2004). Problem-based learning: What and how do students learn? *Educational Psychology Review, 16*(3), 235–266.

Johnson, C. C. (2012). Implementation of STEM education policy: Challenges, progress, and lessons learned. *School Science and Mathematics, 112*(1), 45–55. doi: 10.1111/j.1949-8594.2011.00110.x.

Jones, M. H., Ennes, M., Weedfall, D., Chesnutt, K., & Cayton, E. (2020). The development and validation of a measure of science capital, habitus, and future science interests. *Research in Science Education*. doi: 10.1007/s11165-020-09916-y.

Lent, R. W., Brown, S. D., & Larkin, K. C. (1984). Relation of self-efficacy expectations to academic achievement and persistence. *Journal of Counseling Psychology*, *31*, 356–362.

Lent, R. W., Brown, S. D., Schmidt, J., Brenner, B., Lyons, H., & Treistman, D. (2003). Relation of contextual supports and barriers to choice behavior in engineering majors: test of alternative social cognitive models. *Journal of Counseling Psychology*, *50*(4), 458–465.

Philomin, L. E. (2015). *Science, Math Skills Critical to Singapore's Future: PM Lee*. Retrieved on March 7, 2021 from: https://www.todayonline.com/singapore/sutd-must-champion-science-and-technology-spore-pm-lee.

Renninger, K. A. (2000). Individual interest and its implications for understanding intrinsic motivation. In C. Sansone, & J. M. Harackiewicz (Eds.), *Intrinsic and Extrinsic Motivation* (pp. 373–404). Academic Press.

Shavelson, R. J., Hubner, J. J., & Stanton, G. C. (1976). Self-concept: Validation of construct interpretations. *Review of Educational Research*, *46*(3), 407–441.

Science Centre. (n.d.) *About Our Applied Learning Programme*. Retrieved from: https://www.science.edu.sg/stem-inc/applied-learning-programme/about-our-applied-learning-programme.

Tan, O. S. (2004). Problem-based Learning: The Future Frontiers. In Tan, K., Lee, M., Mok, J. & Ravindran, R. (Eds.). Problem-based Learning: New directions and approaches (p. 17–32). Singapore: Learning Academy, Temasek Centre for Problem-based Learning.

Tan, A. L.*, Teo, T. W., Choy, B. H., & Ong, Y. S. (2019). The S-T-E-M quartet. *Innovation and Education*, *1*(3), 1–14.

Teo, T. W., & Goh, W. P. J. (2019). Assessing lower track students' learning in science inference skills in Singapore. *Asia-Pacific Science Education*, *5*, doi: 10.1186/s41029-019-.

Wenger, E. (1998). Communities of practice: Learning, meaning and identity. *Journal of Mathematics Teacher Education*, *6*, 185–194.

Wilson, Z, S., Holmes, L., Sylvain, M, R., Batiste, L., Johnson, M., McGuire, S, Y., & Warner, I, M. (2012). Hierarchical mentoring: A transformative strategy for improving diversity and retention in undergraduate STEM disciplines. *Journal of Science Education and Technology*, *21*(1), 148–156.

Wigfield, A., & Eccles, J. S. (2000). Expectancy – value theory of achievement motivation. *Contemporary Educational Psychology*, *25*(1), 68–81.

Wigfield, A., & Eccles, J. S. (2002). The development of competence beliefs, expectancies for success, and achievement values from childhood through adolescence. In *Development of achievement motivation*. UK: Academic Press (pp. 91–120).

Appendix: Person and item separation reliabilities

Person and item separation reliabilities for Construct A

Construct A.xlsx

	136 Input		136 Measured		Infit		Outfit	
Person	*Total*	*Count*	*Measure*	*Realse*	*IMNSQ*	*ZSTD*	*OMNSQ*	*ZSTD*
Mean	37.0	10.0	1.11	.50	1.05	−.1	1.00	−.2
S.D.	4.3	.0	.87	.14	.69	1.5	.66	1.4
Real RMSE .52		True SD .70			Separation 1.35		Person Reliability .65	

	10 Input		10 Measured		Infit		Outfit	
Item	*Total*	*Count*	*Measure*	*Realse*	*IMNSQ*	*ZSTD*	*OMNSQ*	*ZSTD*
Mean	503.1	136.0	.00	.13	.98	−.6	1.00	−.4
S.D.	75.9	.0	.95	.01	.46	3.6	.49	3.7
Real RMSE .13		True SD .94			Separation 7.29		Item Reliability .98	

Person and item separation reliabilities for Construct B

Construct B.xlsx

	150 Input		137 Measured		Infit		Outfit	
Person	*Total*	*Count*	*Measure*	*Realse*	*IMNSQ*	*ZSTD*	*OMNSQ*	*ZSTD*
Mean	37.1	10.0	1.37	.55	1.07	.0	1.00	−.2
S.D.	5.2	.0	1.23	.19	.73	1.5	.72	1.4
Real RMSE .58		True SD 1.09			Separation 1.87		Person Reliability .78	

	10 Input		10 Measured		Infit		Outfit	
Item	*Total*	*Count*	*Measure*	*Realse*	*IMNSQ*	*ZSTD*	*OMNSQ*	*ZSTD*
Mean	508.0	137.0	.00	.14	.98	−.6	1.00	−.6
S.D.	66.4	.0	.97	.02	.50	3.6	.57	3.8
Real RMSE .14		True SD .96			Separation 7.00		Item Reliability .98	

Table 8.3 Person and item separation reliabilities for Construct C

Construct C.xlsx

Person	150 Input		134 Measured		Infit		Outfit	
	Total	Count	Measure	Realse	IMNSQ	ZSTD	OMNSQ	ZSTD
Mean	33.0	10.0	.54	.55	1.03	−.3	1.05	−.3
S.D.	6.0	.0	1.42	.17	1.09	1.7	1.19	1.7
Real RMSE .57		True SD 1.30		Separation 2.26		Person Reliability .84		

Item	10 Input		10 Measured		Infit		Outfit	
	Total	Count	Measure	Realse	IMNSQ	ZSTD	OMNSQ	ZSTD
Mean	442.0	134.0	.00	.14	.98	−.5	1.05	−.1
S.D.	49.1	.0	.85	.02	.39	3.0	.49	3.5
Real RMSE .14		True SD .84		Separation 5.94		Item Reliability .97		

Table 8.4 Person and item separation reliabilities for Construct D

Construct D.xlsx

Person	150 Input		135 Measured		Infit		Outfit	
	Total	Count	Measure	Realse	IMNSQ	ZSTD	OMNSQ	ZSTD
Mean	33.0	10.0	.59	.56	1.00	−.4	1.00	−.4
S.D.	6.0	.0	1.54	.19	1.12	1.9	1.13	1.9
Real RMSE .59		True SD 1.42		Separation 2.40		Person Reliability .85		

Item	10 Input		10 Measured		Infit		Outfit	
	Total	Count	Measure	Realse	IMNSQ	ZSTD	OMNSQ	ZSTD
Mean	448.6	135.0	.00	.14	1.00	−.2	1.00	−.1
S.D.	27.0	.0	.47	.01	.27	2.2	.29	2.3
Real RMSE .14		True SD .44		Separation 3.19		Item Reliability .91		

Table 8.5 Person and item separation reliabilities for Construct E

Construct E.xlsx

Person	150 Input		131 Measured		Infit		Outfit	
	Total	Count	Measure	Realse	IMNSQ	ZSTD	OMNSQ	ZSTD
Mean	32.3	10.0	.36	.49	1.00	−.3	1.02	−.3
S.D.	5.8	.0	1.11	.12	.91	1.8	.98	1.9
Real RMSE .50		True SD .99		Separation 1.96		Person Reliability .79		

Item	10 Input		10 Measured		Infit		Outfit	
	Total	Count	Measure	Realse	IMNSQ	ZSTD	OMNSQ	ZSTD
Mean	422.6	131.0	.00	.12	.99	−.2	1.02	.0
S.D.	35.3	.0	.50	.01	.26	2.1	.31	2.3
Real RMSE .12		True SD .48		Separation 3.87		Item Reliability .94		

9 Effect of 5E learning cycle to enhance grade twelve students' conceptual understanding of gene expression

An integrated STEM education approach

Sherab Tenzin, Kinley, and Tempa Wangchuk

Introduction

Bhutan has stepped into the world of science since the introduction of modern education 32 years ago (Jamtsho, 2018). The Bhutanese education journey has scaled many milestones through periodic efforts in the form of policy changes and adaptations to modernistic approaches (Ministry of Education [MoE], 2011a). The modern education journey began with a few hundred students in the early 1960s (MoE, 2014).

With the advent of modern education in Bhutan, science education was introduced into the system. In 1986, the "New Approach to Primary Education" (NAPE) was launched, seeking to orientate the primary science curriculum for Classes IV to VI more confidently, to take account of Bhutanese context, and to uphold the teaching of science-based on Bhutan's natural and social environment (MoE, 2014). The basic elements of science and technology education are included in Environmental Studies (ES) taught in Dzongkha from Pre-primary (PP) to Grade 3 (key stage I) since 1994 and General Science from Grade 4–8. Before 1994, science was taught as an integrated subject from Grade 4–6, and branched into Physics, Chemistry, and Biology disciplines for higher grade levels. However, in 1999 and 2000, the teaching of science in Grades 7–8 was bifurcated into three distinctive science disciplines. Later, a single integrated science of Biology, Chemistry, and Physics, replaced "Science for Class VII and VIII". The science for key stages 2 (Grade 4–5) and 3 (Grade 7–8) was taught as General Science through the integration of concepts of Physics, Biology, and Chemistry (MoE, 2014).

Over the last 15 years, there has been a growing concern both in the Ministry of Education and among international agencies, e.g., World Health Organization, UNESCO, etc. about the quality of education in Bhutan (MoE, 2017). The World Bank (2009) reported low learning levels in competencies to achieve, and a lack of critical thinking, communication, and problem-solving skills among Bhutanese students. Likewise, the Royal Education Council

DOI: 10.4324/9781003099888-9

and Education Initiatives (REC) (2010) revealed that students performed below the expectations of their grade level for both basic and advanced academic skills, and lacked communication and critical skills. The major contributing factor to students' low level of learning has been attributed to ineffective teaching practices (REC, 2019). It was found that frontal teaching method is the dominant teaching approach used in most schools. Classroom teachings were facilitated in traditional ways where teachers take central and dominant roles, and the transmission of knowledge is assumed to have taken place (REC, 2019). Teachers hardly use innovative teaching strategies (e.g., using technology) and constructivist instructional practices in their teaching. Consequently, students hardly have opportunities to actively engage in thinking, sharing ideas, asking questions, and innovative work (Rabgay, 2018). Such learning has led to poor conceptual understanding and a low level of academic achievements (REC, 2019). It was observed that students were passive learners relying on teachers to decide what, when, and how to learn (Rabgay, 2018). In view of this and the current impetus to improve the quality of education in the country, it is imperative for the Bhutanese classrooms to shift from teacher-centred teaching to learner-centred teaching.

Biology lessons in Bhutan

Most biology concepts are taught as an abstract subject in schools without much emphasis on learning strategies and Science, Technology, Engineering, and Mathematics (STEM) integration. This has, in part, resulted in the low acquisition of biology process skills. It has been observed that most students try to memorise and repeat the terms and concepts to pass examinations and would forget most of what they learnt after the examinations (Bahar, 2002). Specifically, genetics and its related concepts in biology have been considered as the most difficult concepts to learn by students (Bahar, 2002). In the last century, however, international researchers have offered alternative strategies for the meaningful learning of biology concepts such as using various information and communications technology (ICT) tools for active learning. Therefore, learning biology can be made easier, enjoyable, and more motivating by integrating ICT tools in the instructional teaching strategies for biology.

The 5E learning cycle instruction (LCI) is widely used today in different countries as an integral component of many teaching practices and research attempts to enhance students' outcomes (Marek et al., 2003). The 5E learning cycle has been used since the late 1980s when it was proposed by Biological Science Curriculum Study (BSCS) under the theories of the constructivist teaching model (Bybee, Taylor, Gardner, et al., 2006). Research findings suggests that 5E LCI could improve reasoning skills (Schnieder & Renner, 1980), conceptual achievement (Balci, 2009), scientific attitudes (Oren & Tezcan, 2009), and bring about conceptual change (Bybee et al., 2006). Further, the 5E learning cycle is the most effective strategy for teaching a difficult concept of science. The use of the 5E model in the classroom helps to facilitate inquiry practices, focusing on constructivist principles and emphasising the explanation and investigation of phenomena, the use of evidence to back up conclusions, and experimental design (Bybee, Taylor, Gardner, et al., 2006). The 5E model also serves as a flexible learning cycle to assist curriculum developers and classroom teachers to create science lessons. Yalcin and Bayrakceken (2010) applied the 5E learning cycle and found that activities based on the 5Es learning cycle are efficient and resourceful in science classes to improve students learning meaningfully.

Studies on the use of the 5E learning cycle approach to teach genetics in Bhutanese higher secondary schools is limited. As such, this study aimed to investigate the effectiveness of the 5Es learning cycle in the Bhutanese curricula to enhance students' conceptual understanding in science, incorporating an integrated STEM Education approach. The study has drawn inspiration from the framework on integrated STEM education (Bryan et al., 2015). The conceptual clarification of Tan et al. (2019) on integrated STEM education approach "putting problem solving at the heart of integrated STEM education potentially provides an authentic and meaningful context that motivates learning by bringing together individual disciplines, capitalising on the knowledge and skills of each discipline to solve the complex problem" (2019, p. 6). In this study, the delivery of a biology concept through the 5E learning and research process was done with an integrated STEM Education approach utilising mathematics and technology to provide meaningful learning and gather insightful research findings. Mathematics has been fundamental in the conceptual development of the science of genomic sequences, for describing populations, and key for pattern recognition in inheritance. The teaching of quantitative biology has attempted to connect mathematical reasoning with biological understanding. In probability and statistical analyses, mathematical concepts are commonly used. Students often assume that learning Biology is difficult due to the theoretical and abstract nature of the subject. Therefore, this study attempted to make learning biology more interesting and meaningful by integrating ICT tools in the instructional strategies for teaching biology. Visual presentations, simulations, and animations of the concept of genetics were used to make learning easier.

Research questions

1 What is the effect of the 5E learning cycle on Grade 12 Bhutanese students' learning achievements in genetics?
2 What is the effect of the 5E learning model on Grade 12 Bhutanese students' opinion towards genetics as indicated by their level of understanding, interest, satisfaction, and difficulty in learning?
3 What effect do the five phases of the 5E learning cycle have on teaching and learning after incorporating the integrated STEM Education Approach?

Research hypothesis

1 **H₀ (Null hypothesis):** There is no significant effect of the 5E learning cycle on students' learning achievement.
2 **Hₐ (Alternative hypothesis):** There is a significant effect of the 5E learning cycle on students' learning achievement.

Literature review

This section presents the review of literature that includes the definition of the 5E learning cycle, phases of the 5E learning cycle, theories underlying the 5E learning cycle, and the effects of the 5E learning cycle.

The 5E learning model concept

The 5E learning cycle is a systematic model of inquiry instruction. It is regarded as a universal philosophy of teaching and learning with well-built constructivist foundations (Cornelius, 2012). In the same way, the 5Es learning cycle also helps in conceptual change by letting students "redefine, reorganise, elaborate, and change their prior concepts through interaction and self-reflection with their peers and their environment" (Bybee, 1997). According to Duran and Duran (2004), the 5E learning cycle helps and serves teachers by providing the approach in course processing and good reform-based instruction to develop the curriculum. Besides, the technology-enriched 5E learning cycle is a first-class tool for learners to obtain 21st-century skills and for the teachers to teach a precise concepts (Senan, 2013). Therefore, the five stages of the 5E learning model can be embedded with all aspects of inquiry learning situations to create opportunities for students to engage and explore the concepts, to craft explanations for the concepts learned, to elaborate by applying their prior knowledge to new situations, and to provide an evaluation for students' understanding (Orgil & Thomas, 2007).

Traits of the 5E learning cycle approach

The five stages of the 5E learning cycle are engagement, exploration, explanation, elaboration, and evaluation.

Engagement

In the first phase, students encounter and identify the instructional task, and it creates students' interest in the subject by generating curiosity, raising questions, and eliciting thoughts and responses that unearth prior knowledge to discover student preconceptions (Lynn, 2012). Therefore, this stage is also called a warm-up phase as students are getting prepared to learn and hence is perceived to be the most critical phase. The role of teachers in this phase is to obtain prior knowledge and get the students interested in the topic of study. It is the most active phase of the 5E model for the learners.

Exploration

In the second phase, students "have common, concrete experiences upon which they continue building concepts, processes, and skills" (Bybee, 1997, p. 51), and students use their prior knowledge to observe and investigate to explore the scientific knowledge (Cepni et al., 2010). Additionally, working on group activities provides students with concepts and skills to help them use prior knowledge to generate new ideas and help explore new possibilities (Cornelius, 2012; Crider, 2013). Finally, students and teachers use their experiences from this phase to make meaning of current concepts and processes, skills are identified, and conceptual change is facilitated for explanation in the next phase.

Explanation

The root definition of the explanation phase stated by Bybee is "to present concepts, processes, or skills briefly, simply, clearly, and directly" (1997, p. 52). This phase provides students with an opportunity to communicate what they have learned and figure out what it means.

Elaboration

This phase encourages students to widen their understanding of a scientific concept, what they have experienced through the earlier three stages (Cornelius, 2012). The primary goal of the phase is to generalise concepts, processes, and skills (Bybee, 1997). Teachers make students use new terminology and encourage them to apply their new skills to different situations. Therefore, students enlarge on the concepts they have learned, make connections to other related concepts, and apply their understandings to the world around them through additional activities.

Evaluation

The evaluation phase is a critical one. It allows for students to conduct self-assessment or peer-assessment, and allows teachers to examine student learning as they apply new concepts and skills, look for evidence that students have changed or modified their thinking or behaviour, and evaluate for student misunderstanding (Bybee, 2009; Bybee et al., 2006). In practice, this phase should be embedded in all phases of formative assessment during the inquiry-based lesson. Assessment should be viewed as an ongoing process (Duran & Duran, 2004; Lynn, 2012).

Effects of the 5E learning cycle approach

The 5Es learning cycle is an important instructional model in teaching and learning of science, as it is able to motivate students to think creatively and critically, facilitate a better understanding of scientific concepts, develop a positive attitude towards science and higher-order thinking skills, and cultivate advanced reasoning skills over conventional instruction (Arslan, 2014). The 5E learning cycle is an effective method in bringing conceptual change or removing misconceptions in the teaching of biology concepts such as the circulatory system, cell concepts and its organelles (Urey & Calik, 2008), cell cycle and also enhance retention (Fazelian et al., 2010), as well as improving the attitudes and interest towards learning of science (Bybee et al., 2006).

Hagerman's (2012) research investigated the effectiveness of the 5E LCI on 42 twelfth-grade students' achievements in three units – cellular structure, genetics, and evolution. A sample size of 42 students was chosen to reveal significant effectiveness. The 5E LCI is not only used in science education but it is also found effective in mathematics education and other disciplines in different countries (Cakır, 2017). Further, studies have also been carried out in Italy, Turkey, Nigeria, and America on the effectiveness of the 5E learning cycle in different disciplines. It was reported that the 5E learning cycle enhanced students' learning achievements compared to normal teaching (Temel et al., 2013).

STEM education in Bhutan

A study by Childs et al. (2012) contended that some of the issues and challenges pertaining to science education in Bhutan included inadequate content and pedagogical knowledge of teachers, and fragmentation and discontinuity in the current science curriculum. Science and mathematics are compulsory subject for children and the curriculum forces students to study it, despite low levels of science and mathematical understanding and lack of interest in mathematics (Rinzin, 2018). Further, the integration of ICT in teaching and learning processes to promote interest in learning is minimal. Barriers such as a lack of ICT resources hinder successful integration of ICT for teaching and learning in Bhutanese schools. Hence, there is a need to change the way mathematics and science are currently taught in the schools and colleges of Bhutan. Students' fluency in mathematics and science is needed to help solve the challenges faced by Bhutan, by building a scientifically literate society, enhancing the country's economy, and mitigating employment problems. The S-T-E-M Quartet instructional framework provides inspiration to help teachers think about how the different disciplines are connected and a pathway towards an integrated approach to STEM education (Tan et al., 2019). Therefore, it is inspiring to use integrated STEM approach for enhancing learning following the explanation of Moore and Smith (2014, p. 5) "integrated STEM education is an effort to combine the four disciplines of science, technology, engineering, and mathematics into one class, unit, or lesson that is based on connections among these disciplines and real-world problems". The country aims to enhance students learning achievements in STEM subjects to build human capital with adequate scientific attitude and skills to facilitate its development process (Rabgay, 2018). While Bhutan has made progressive attempts in enhancing science education through the infusion of STEM education concepts, no significant achievements have been made in the area of integrated STEM education. Therefore, this study will be a pioneer in presenting some initial steps in the application of integrated STEM education efforts.

Method

The purpose of this study was to investigate the effectiveness of the 5E learning cycle on the mean achievement scores of twelfth-grade students on the concept of genetics and their opinions towards the concept in terms of the level of interest, understanding, satisfaction, and difficulty. Additionally, we also focused on students' opinions towards teaching and learning through the incorporation of the 5E learning model. Therefore, the study used a mixed-method research approach. The qualitative component included students' opinions on the 5E learning cycle collected through questionnaires, and classroom observation checklists built on a four-point Likert scale. The quantitative component included an experimental design where the 5E learning cycle was the independent variable while students' test scores and opinions towards Biology were the dependent variables.

Participants

The participants of this study were 61 Grade 12 students of Biology and three biology teachers were drawn from three selected higher secondary schools from Mongar district in eastern Bhutan, as shown in Table 9.1.

In the study, participants were divided into two groups, the experimental group (EG) and the control group (CG). A pre-test using the Gene Expression Learning Achievement Test (GELAT) was administered to both the CG and EG students. To ascertain the comparability of the two groups in terms of academic competence, sampled students were divided equally based on their pre-test

Table 9.1 Demographics of the students in the three schools

| School type | Section | Gender | | Total |
		Boy	Girl	
Yadi Central School (Standard PP–12)	I	12	11	23
Mongar Higher Secondary School (Standard 9–12)	I	13	8	21
Sherab Reldi Higher Secondary School (Standard 11–12)	I	8	9	17
Total		34	28	61

scores. The CG was taught using the Traditional Teaching Method (TTM) whilst the EG was taught using the 5E learning cycle on the concept of genetics. In the EG, the lessons were student-centred, students were free to express their ideas and they were actively engaged with visual presentations, animations, and simulations. However, in the CG, the lessons involved frontal teaching and students did not get to share their ideas in the classroom.

Curriculum

Before the intervention, teachers for the EG was trained to use the 5E learning model combined with STEM for about two days. At the end of the training, the teachers for the EG were given a copy of the developed instructional lesson. The training session included a segment to familiarise the teachers with teaching materials. Subsequently, the student participants in the EG were first trained by their teacher during the first few classes on the 5E learning cycle as students have no prior experience of learning in 5E learning cycle groups. After the training, the teachers were deployed to teach the EG using the 5E learning cycle method and the researcher/teacher observer served as a participatory observer of all the lessons in the EG. The GELAT post-test was administered to measure the effectiveness of the 5E learning cycle combined with STEM after the intervention. Both the groups were taught for 200 minutes. Figure 9.1 below provides an overview of this study:

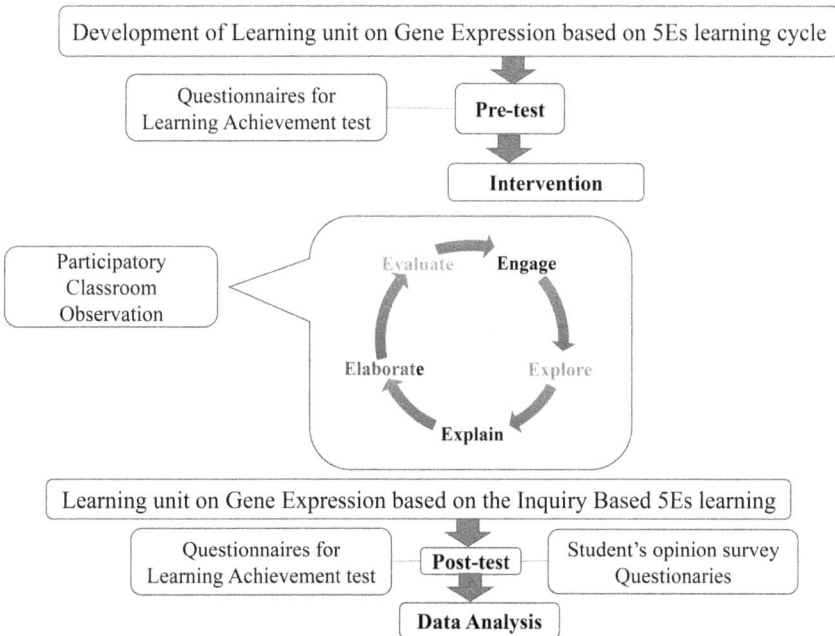

Figure 9.1 The research design of the study (adapted from Mustafa, 2016).

In the EG the integration of STEM in 5E biology lesson involved STEM-related classroom practices such as student-centered activities (e.g., hands-on activities building models of DNA and RNA, to games and presentations) and the use of technology (online videos explaining concepts that are abstract, and animations depicting transcription and translation to synthesis protein. On the other hand, in the CG students were taught the same concept through conventional method using textbook and chalk board.

Research instrument

In the study, four instruments were used to gather data: (1) GELAT, (2) Student Opinion Survey Questionnaires (SOSQ), (3) Containing closed- and open-ended questions, and (4) Participatory Classroom Observation Checklist (PCOC). The learning achievement test consisted of 20 multiple-choice questions with a maximum score of 20 based on "Gene Expression", which was used in both groups.

To ascertain the content validity of the test questions, past question papers prepared by the Bhutan Council of School Examination and Assessment (BCSEA) were used to assess learners on competencies and standards required in biology to achieve content validity. In addition, one biology curriculum officer, one biology senior teacher, and one research officer who were familiar with the biology content evaluated the test questions and provided expert validation. They ascertained that the test items were based on the content and specific objectives of "Gene Expression" as prescribed in the Bhutanese curriculum's school biology syllabus. To ascertain the reliability of the instrument, the test was pilot-tested in another school, Lhuntse Higher Secondary School, which was not a part of the study. To test the internal consistency reliability, we analysed the data for Cronbach's alpha and obtained a reliability coefficient of 0.66 (N = 22). Therefore, the GELAT was a reliable instrument, and the scores obtained could be used in the study.

The survey questionnaires consisted of 16 statements (closed-ended questions) and two open-ended questions related to students' opinions towards the concept and instruction based on the 5E learning cycle approach. The SOSQ was built using a four-point Likert scale (4 = *Strongly Agree (SA)*, 3 = *Agree (A)*, 2 = *Disagree (DA)*, and 1 = *Strongly Disagree (SDA)*). A reliability coefficient of 0.60 (N = 31) was obtained. Therefore, the SOSQ was a reliable instrument to use in the study. Further expert validation was done in consultation with one biology curriculum officer and one research officer who is experienced in educational research. It was mainly to check and validate the Item Objective Congruent (IOC). After making a few corrections, they confirmed that the questionnaires were valid for use in the study.

The PCOC was prepared based on the five phases of the 5E learning cycle. The purpose of the lesson observation checklist was to study the level to which the lessons were characterised by the 5E learning environment. To ascertain the validity the statements, they were shown to one biology curriculum officer and one research officer. They confirmed that statements were valid for use in the field. The PCOC was used to observe 12 5E learning cycle lessons of the experimental group in the three schools with the integrated STEM education approach.

Data analysis approach

Data analysis was conducted on the test scores, students' opinions, and classroom observation checklist. For the test score and students' opinion analysis, means, and standard deviations were computed. A paired-samples t-test was used to determine the difference in test scores and change in opinion. The PCOC analysis was done according to the five phases of the 5E learning cycle. Means and standard deviations were computed for each phase. Each mean represented a level of opinion on the scale which indicated whether the stages were followed during the lesson.

Results and discussion

Students' GELA test score analysis

The results of test score analysis showed that the pre-test means of the CG and EG were not different but on post-test, the mean of the EG was significantly higher than that of the CG. The analysis is shown in Table 9.2. The mean difference of post-test in the EG and CG is 5.64 (14.74–9.10) and the mean difference of pre-test in EG and CG is 0.31 (8.94–8.63). The mean difference indicated that there was significantly high score difference between the post-test as compared to the pre-test. It is also supported by value of Cohen's d, with 4.99 indicating a large effect in post-test, and with 0.27 indicating small effect in the pre-test. Moreover, a significant difference was found between the pre-test and post-test in the EG and CG ($p < 0.05$). With this result, the null hypothesis was therefore rejected because there was significant effect of the 5E learning cycle on students' learning achievement.

The results of the study revealed that the effect of 5E LCI on students' post achievement scores on gene expression was statistically significant ($p > 0.05$) compared to Conventional Classroom Instruction (CCI) students. In addition,

Table 9.2 Comparison of pre-post-test of GELA scores between the groups (EG and CG)

Test	Group	Mean	Mean difference	Standard deviation	Sig. (2 tailed)	Cohen's d
Post-test	Experiment	14.74	5.64	1.41	0.00	4.99
	Control	9.10		1.12		
Pre-test	Experiment	8.94	0.31	1.00	0.31	0.27
	Control	8.63		1.27		

Significant level: >0.05 – not significant, <0.05 – significant.

Cohen's *d* value: $d = 0.2$ – small effect, $d = 0.5$ – medium effect, $d = 0.8$ – large effect.

the estimated Cohen's d value of post-test between EG and CG was 4.99, which indicated that there was a large practical significant effect on student's GELA scores between the EG and CG. On the other hand, Cohen's d value of pre-test between EG and CG was 0.27, which indicated that there was a small practical significant effect in students' GELA scores between the EG and CG.

The 5E learning cycle facilitated effective learning by creating a socially oriented learning environment (Vygotsky, 1978). Such a learning environment in the EG enabled students' interaction, the free share of ideas, and support for each other in their learning process (Johnson & Johnson, 1990).

The students who were exposed to 5E LCI spent more time on different types of hands-on-activities and minds-on-activities embedded with real world-based learning units designed in the study, whereby learning biology was made easier and more comfortable by integrating ICT tools in instructional teaching strategies. Students in the CCI group were in a passive position listening to their teacher, following textbooks, and taking notes most of the class time. In this kind of atmosphere, students did not have the chance to share their ideas with their friends and teachers.

Based on ideas from Pulat's (2009) study, it is likely that the students in the CG found the lessons boring, while the students in EG found the lessons interesting and had a lot of fun. Therefore, the GELAT scores difference between pre-test and post-test was not due to chance but is likely to be attributed to the treatment. Our findings aligned with Vygotsky's (1978) notion that learning is facilitated by social interaction, with more experienced individuals guiding the learning process. The findings also concur with a study by Bello (2012), regarding the promising effects of the 5E learning cycle towards enhancing students' conceptual understanding of genetics. Therefore, the triangulation of the quantitative and qualitative results from open-ended questions of this study revealed that the EG students constructed more meaningful learning than the CG. Further, the findings are also in agreement with studies conducted in different disciplines other than biology with significant gains in the learning achievements of students in the EG than students in the CG (Anil & Batdi, 2015; Bulbul, 2010; Sarac, 2017; Ural & Bumen, 2016). The study conducted based on the effectiveness of the 5E learning model in Bhutan also revealed that the incorporation of 5E learning cycle method in the teaching of science offers learners the opportunity for active learning in contrast to the overriding conventional teacher-centred method of teaching and learning approach (Pelzang, 2012). It brought about a paradigm shift from the traditional classroom into an active classroom with enhanced learning experiences in biology by leveraging technology.

Students' opinion analysis

Students' opinion analysis of the EG presented in Table 9.3 showed that the pre-survey means of interest, understanding, and satisfaction in learning the gene expression concept increased to higher levels in the post-survey. The stage of

Table 9.3 Comparison of students' opinion pre-post-survey of EG

Opinion	Post-survey			Pre-survey		
	Mean	*SD*	*Level of opinion*	*Mean*	*SD*	*Level of opinion*
Interest	3.77	0.70	Very high	2.39	0.54	High
Understanding	3.65	0.71	Very high	2.86	0.58	High
Satisfaction	3.69	0.76	Very high	2.97	0.54	High
Difficulty	1.31	0.40	Very low	3.90	0.70	Very High

Level of opinion: 1–1.75: Very low, 1.76–2.50: Low, 2.51–3.25: High, 3.26–4.00: Very high.

complexity of the subject reduced from very high to very low in the pre-post sur-vey respectively. The results signify that students' level of interest, understanding, and satisfaction in learning the concept of genetic elevated and indicated that the concept was less difficult. It indicated that students exposed to 5E learning instruction found concept interesting, understanding, satisfactory and had less difficulty understanding the concept.

Our findings exhibited that there was an improvement in student's opinions towards biology as indicated by their increased level of interest, understanding, satisfaction, and their assessment of the concept of genetics as a less difficult topic. These outcomes paralleled those of Hiccan's (2008) that instructions based on the 5E learning cycle increased the seventh-grade students' interest, motivations, and participation towards the subject. It was also consistent with the findings of Yilmaz et al. (2011) who examined the effect of the 5E learn-ing cycle on students' achievement and attitude towards biology and concluded that the 5E learning model increased students' achievement and promoted pos-itive opinion towards studying biology. Our findings were also congruent with the claim that the 5E learning model was more effective on students' academic achievement, attitude towards the lesson and science process skills than the con-ventional teaching method (Cakır, 2017). Further, our findings were coherent with the findings of Namgyel and Buaraphan (2017) who revealed that the stu-dents' opinions towards science were higher in the case of students who were exposed to the LCI method than those students who were exposed to the CCI. On this note, it was also revealed that 5E LCI enhanced not only the 10th-grade Bhutanese students' understanding of the DNA and chromosome structure con-cepts but also motivated students, and boosted positive attitudes towards the learning unit (Pelzang, 2012).

The prevalence of increased levels of students' interest and satisfaction could be due to the enjoyable learning environment created by the various activities based on the five phases of the learning cycle inbuilt with ICT tools, such as games, animation, etc. The students' positive attitude might have been a result of the phase-wise activities integration of technology, teacher's approach, atti-tude, and beliefs. However, it also depicted that students need a teacher with strong content mastery and keen understanding of learners' needs, one who is not annoying, and who teaches the lesson at the level of the student. Thus, teachers need to work with the interests of students, their abilities, and personalities besides the lessons taught (Pulat, 2009). Likewise, students also expressed that when the PowerPoint presentations, animation, games, and materials used during the lesson were interesting and entertaining, their motivation and concerns about the subject increased. However, normally in Bhutan, teachers hardly use technology, materials, and various activities that make class more lively and interesting; teachers are authoritative figures in the class and students hardly get the opportunity to share their ideas (Rabgay, 2018).

Classroom observation checklist analysis

The results of classroom observation analysis presented in Table 9.4 depicted that teacher observers actively settled that the 5E learning model was characterised by five phases of the STEM education approach – engagement, exploration, explanation, elaboration, and evaluation. The overall mean of five phases is 3.66. It indicated that the level of opinion for five phases was "strongly agree", whereby, students demonstrated various skills individually as well as in groups during the 5E learning cycle's activities that incorporated various ICT tools in each phase. The use of ICT tools in the 5E activities made it easier for understanding the concept of genetics, as it simplified the teaching of the concept with the use of visual presentations. Therefore, through new learning experiences such as visual presentations, simulations, and animations, the concept of genetics could be easily understood by students. It showed that the lessons in the EG were carried out in 5E learning cycle environments using various ICT tools.

The increase in student's achievement test scores and improvement in their opinion towards the concept of genetics was attributed to the five phases of the 5E learning cycle, which incorporated characteristics such as a social learning context, a non-threatening classroom, equal learning opportunity, and a student-centred learning approach to construct their knowledge, self-learning evaluation and discussions, reinforcing retention in memory through various technology tools. The phases of the 5E model enable not only students to explore ICT but also practitioners to become familiar with new technology tools. According to the findings of the open-ended questions in the study, it was

Table 9.4 Comparison of means and opinion levels of teachers' observation during the lesson

Phases of the 5E model	Mean	SD	Level of opinion
Engagement	3.46	0.34	Strongly agree
Exploration	3.67	0.24	Strongly agree
Explanation	3.75	0.17	Strongly agree
Elaboration	3.67	0.24	Strongly agree
Evaluation	3.75	0.17	Strongly agree

Level of opinion: 1–1.75: strongly disagree; 1.76–2.50: disagree; 2.51–3.25: agree; 3.26–4.00: strongly agree.

depicted that learning a lesson with 5E learning instruction was better than just learning through the book and it helped students to grasp the topics, enhance their understanding, and facilitate the reconstruction of their existing knowledge. Besides, students' active participation and free interaction at every stage of the 5E learning cycle in the EG were evident from the open-ended question analysis and teacher observer's notes. In the same vein, the 5E learning cycle with integration of ICT tools created an atmosphere in which students felt free to express their ideas and there was no time to get bored.

Conclusion

This study found that the 5E learning cycle approach promoted the higher academic achievement of twelfth-grade Bhutanese Higher Secondary School students in biology as compared to the traditional based teaching method. The approach also had a positive effect on students' opinions towards the concept of gene expression as indicated by their higher level of interest, satisfaction, understanding, and their assessment of the concept as a less difficult subject. Therefore, the benefit of the 5E learning method is not culture-bound but it is also effective in raising students' academic scores and opinions towards biology in the Bhutanese school context. The findings of this study exemplified the application of integrated STEM education approach through the delivery of a biology lesson with the utilisation of knowledge from mathematics and various technology tools. Finally, the integrated approach was also outlined through the research process.

References

Anil, V., & Batdi, V. (2015). A comparative meta-analysis of 5E and traditional approaches in Turkey. *Journal of Education and Training Studies, 3*(6), 212–219. doi: 10.11114/jets.v3i6.1038.

Arslan, H. Ö. (2014). *The Effect of 5E Learning Cycle Instruction on 10th Grade Students' Understanding of Cell Division and Reproduction Concepts*. Unpublished thesis for the Master, Middle East Technical University, Turkey.

Bahar, M. (2002). Students' learning difficulties in biology: Reasons and solutions. *Journal of Kastamonu Faculty of Education, 10,* 73–82.

Balci, S. (2009). *The Effects of 5E Learning Cycle Model Based on Constructivist Theory on the Academic Success of Students in Biology Education*. Unpublished Master's thesis. Gazi University, Ankara, Turkey.

Bank, W. (2009). *Findings from the Bhutan Learning Quality Survey*. South Asia human development sector series; no. 21. World Bank.

Bello, A. (2012). *Effects of Learning Cycle Teaching Strategy on Students' Acquisition of Formal Reasoning Ability and Academic Achievement in Genetics in SSII Biology Students in Sabon-Gari, Kaduna State*. Unpublished PhD (Science Education) Dissertation, Ahmadu Bello University.

Bulbul, Y. (2010). *Effects of Learning Cycle Model Accompanied with Computer Animations on Understanding of Diffusion and Osmosis Concepts*. Unpublished doctoral dissertation, Middle East Technical University, Ankara, Turkey.

Bryan, L. A., Moore, T. J., Johnson, C. C., & Roehrig, G. H. (2015). Integrated STEM education. *STEM Road Map: A Framework for Integrated STEM Education*, 23–37.

Bybee, R. W. (2009). The BSCS 5E instructional model and 21st century skills. *Biological Science Curriculum Study, 24.*

Bybee, R., Taylor, J. A., Gardner, A., Scotter, P. V., Powell, J. C., Westbrook, A., & Landes, N. (2006). The basic 5E instructional model: Origins and effectiveness. *Office of Science Education National Institutes of Health*, 1–8.

Bybee, R. W. (1997). *Achieving Scientific Literacy: From Purposes to Practices*. Westport: Heinemann.

Bybee, W. R., Joseph, A. T., April, G., Pamela, V. S., Janet, C. P., Anne, W., & Nancy, L. (2006). The BSCS 5E instructional model: Origins, effectiveness, and applications. *Colorado Springs, CO: BSCS.*

Cakır, N. K. (2017). Effect of 5E learning model on academic achievement, attitude and science process skills: Meta-analysis study. *Journal of Education and Training Studies*, 5(October), 157–170. doi: 10.11114/jets.v5i11.2649.

Cepni, S., Sahin, C., & Ipek, H. (2010). Teaching of floating and sinking concepts with different methods and techniques based on 5E instructional model. *Asia-Pacific Forum on Science Learning and Teaching*, 11(2), 1–38.

Childs, A., Tenzin, W., Johnson, D., & Ramachandran, K. (2012). Science education in Bhutan: Issues and challenges. *International Journal of Science Education*, 34(3), 375–400, doi: 10.1080/09500693.2011.626461.

Cohen, J. (1988). *Statistical Power Analysis for the Behavioral Sciences* (2nd Ed.). Hilsdale, NJ: Lawrence Erlbaum.

Cornelius, M. (2012). *The 5E Learning Cycle and Students Understanding of the Nature of Science.* MEd thesis. Montana State University Bozeman, Montana.

Crider, J. C. (2013). *The 5E Learning Cycle vs. Traditional Teaching Methods and How They Affect Student Achievement, Interest, and Engagement in a Third Grade Science Classroom.* Thesis. Montana State University Bozeman.

Duran, B. L., & Duran, E. (2004). The 5E instructional model : A learning cycle approach for inquiry-based science teaching. *The Science Education Review*, 3(2), 49–58.

Fazelian, P., Ebrahim, A. N., & Soraghi, S. (2010). The effect of 5E instructional design model on learning and retention of sciences for middle class students. *Procedia – Social and Behavioral Sciences.* doi: 10.1016/j.sbspro.2010.07.062.

Hagerman, C. L. (2012). *Effects of the 5E Learning Cycle on Student Content Comprehension and Scientific Literacy.* Unpublished master's thesis, Montana State University, USA.

Hiccan, B. (2008). *The 5e Effect of Teaching Activities Based on Learning Cycle Model on the Academic Achievements of First Grade 7th Students on the Basis of Mathematics Lesson First Degree Unknowns.* Unpublished Master thesis, Gazi Universities, Institute of Educational Sciences, Ankara.

Jamtsho, S. (2018). *Education in Bhutan: Quality and Sustainability.*

Johnson, D. W., & Johnson, R. T. (1990). Social skills for successful group work. *Educational Leadership*, 47(4), 29–33.

Lynn, H. B. (2012). Guided Inquiry using the 5E Instructional Model with High School Physics. *Montana State University Bozeman, Montana.*

Marek, E. A., Laubach, T. A., & Pedersen, J. (2003). Preservice elementary school teachers' understandings of theory based science education. *Journal of Science Teacher Education*, 14(3), 147–159.

Ministry of Education. (2011a). *Science Curriculum Framework PP–XII.* Ministry of Education. Thimphu, Bhutan.

Ministry of Education. (2014). *Bhutan Education Blueprint*. Ministry of Education. Thimphu, Bhutan.

Ministry of Education. (2017). 31st educational policy guidelines and instructions (EPGI) 2013–2017. In *Policy and Planning Division Ministry of Education*. Thimphu: Ministry of Education.

Moore, T. J., & Smith, K. A. (2014). Advancing the state of the art of STEM integration. *Journal of STEM Education: Innovations and Research, 15*(1), 5.

Mustafa, M. E. I. (2016). The impact of experiencing 5E learning cycle on developing science teachers technological pedagogical content knowledge (TPACK), *4*(10), 2244–2267. doi: 10.13189/ujer.2016.041003.

Namgyel, T., & Buaraphan, K. (2017). The development of simulation and game in 5E learning cycle to teach photoelectric effect for grade 12 students. *Asia-Pacific Forum on Science Learning and Teaching, 18*(2), 1–31.

Oren, F. S., & Tezcan, R. (2009). The effectiveness of the learning cycle approach on learners' attitude toward science in seventh grade science classes of elementary school. *Elementary Education Online, 8*(1), 103–118.

Orgil, M. K., & Thomas, M. (2007). Analogies and the 5E model, suggestions for using analogies in each phase of the 5E model. *The Science Teacher, 74*(1), 40–45.

Pelzang, T. (2012). *Enhancing Grade 10 Bhutanese Students' Understanding of the DNA and Chromosome Structure Using 5Es Learning Cycle*. A thesis submitted in partial fulfillment of the requirements for the degree of master of science (science and technology education), Faculty of Graduate Studies Mahidol University, Thailand.

Pulat, S. (2009). *Impact of 5E Learning Cycle on Sixth Grade Students' Mathematics Achievement and Attitudes Toward Mathematics*. A thesis submitted to the graduate school of social sciences of Middle East Technical University, Turkey.

Rabgay, T. (2018). The effect of using cooperative learning method on tenth grade students' learning achievement and attitude towards biology, *11*(2), 265–280.

Rinzin, K. (2018, January 15). Education in Bhutan – Achievements and Challenges – Part 1 of 2. *The Bhutanese*.

Royal Education Council (REC). (2019). Annual Report. Royal Education Council. Paro, Bhutan.

Royal Education Council & Education Initiatives. (2010). *Bhutan's Annual Status of Student Learning*. Thimphu: Royal Education Council.

Sarac, H. (2017). The effect of 5E learning model usage on students' learning outcomes: Meta-analysis study. *The Journal of Limitless Education and Research, 2*(2), 16–49.

Schneider, L. S., & Renner, J. W. (1980). Concrete and formal teaching. *Journal of Research in Science Teaching, 17*(6), 503–517.

Senan, D. C. (2013). Infusing BSCS 5E instructional model with multimedia: A promising approach to develop 21st century skills. *I Managers' Journal on School Educational Technology, 9*(22), 1–7.

Singh, S., & Yaduvanshi, S. (2015). Constructivism in science classroom: Why and how. *International Journal of Scientific and Research Publications, 5*(3), 1–5.

Tan, A. L., Teo, T. W., Choy, B. H., & Ong, Y. S. (2019). The STEM quartet. *Innovation and Education, 1*(1), 1–14.

Temel, S., Yılmaz, A., & Özgür, S. D. (2013). Use of the learning cycle model in the teaching of chemical bonding and an investigation of diverse variables in prediction of achievement, *1*(5), 1–14.

Ural, G., & Bumen, N. (2016). A meta-analysis on instructional applications of constructivism in science and technology teaching: A sample of Turkey. *Education and Science,* *41*(185). doi: 10.15390/EB.2016.4289.

Urey, M., & Calik, M. (2008). Combining different conceptual change methods within 5E model: A sample teaching of cell concepts and its organelles. *Asia-Pacific Forum on Science Learning and Teaching, 9*(2), 1–16.

Vygotsky, L. S. (1978). *Mind in Society: The Development of Higher Mental Processes.* Cambridge, MA: Harvard University Press.

Yalcin, F. A., & Bayrakceken, S. (2010). The effect of 5E learning model on pre-service science tearcher' achievement of acids-bases subject. *International Online Journal of Educational Sciences, 2*(2), 508–531.

Yilmaz, D., Tekkaya, C., & Sungur, S. (2011). The comparative effects of prediction/discussion-based learning cycle, conceptual change text, and traditional instructions on student understanding of genetics. *International Journal of Science Education, 33*(5), 607–628.

Section 4

STEM education research in Asia

Envisioning a new education paradigm

Yann Shiou Ong

States of STEM education research in Malaysia, South Korea, and Singapore

The three chapters in this book section provide insights on the states of science, technology, engineering, and mathematics (STEM) education research in three Asian educational systems – Malaysia, South Korea, and Singapore – at different grain sizes. Rohaida Mohd Saat and Hidayah Mohd Fadzil (Chapter 10) offered an overview of STEM education research trends in Malaysia over a decade (year 2010 to 2020). Through their literature review, the authors reported a general increase in the number of STEM education publications over time. The publications included mainly education research in individual STEM disciplines and research in two or more STEM disciplines i.e., integrated STEM. Majority of the STEM education studies in Malaysia were conducted at the secondary level; few studies were conducted at the primary or preschool level. Studies mostly involved students as participants or a mixture of participants (e.g., students and teachers) and research-based activities i.e., interventions rather than outreach programmes (e.g., STEM seminars, industry visits, competitions). The authors' findings correspond to findings from Takeuchi et al.'s (2020) review of trans-disciplinary STEM articles between 2007 and 2018 (which did not include any article representing STEM education in Asia). Similar to Mohd Saat and Mohd Fadzil's conclusions, Takeuchi and colleagues found that their reviewed studies focused on secondary and post-secondary education with few studies in early years and elementary levels (n = 6). Furthermore, most studies were conducted in school/formal education settings and few were in out-of-school settings (n = 15).

Oksu Hong (Chapter 11) summarised the STEAM education research efforts in South Korea aligned with national initiatives in STEAM education and supported by the Korea Foundation for the Advancement of Science and Creativity (KOFAC). Similar to the research efforts in Malaysia, these include education

DOI: 10.4324/9781003099888-IV

research in individual STEM disciplines (i.e., in science education, mathematics education, and informatics education) as well as research in integrated STEAM education. As Hong noted, integrated STEAM education in South Korea is characterised by the keyword, "convergence". One definition of convergence is "the deep integration of knowledge, techniques, and expertise from multiple fields to form new and expanded frameworks for addressing scientific and societal challenges and opportunities" (National Science Foundation, n.d.). A key feature across the Korean STEAM education policies and corresponding research efforts is an emphasis on the use of cutting-edge scientific technologies, including artificial intelligence (AI), augmented reality (AR)/virtual reality (VR), and Internet of things (IoT).

At a finer grain size, Aik-Ling Tan and Roxanne Lau Shu Xin (Chapter 12) described a single case enactment of an integrated STEM activity in a Singapore secondary school classroom. The activity was designed using the S-T-E-M Quartet instructional framework and involves group problem-solving of a complex, persistent, and extended problem. Through engagement in the integrated STEM activity, students could learn the deep or vertical disciplinary knowledge as well as the interconnections across the knowledge in the STEM disciplines i.e., the horizontal connections. Although the authors did not present findings associated with student learning of vertical knowledge and horizontal connections, they highlighted how students' engagement in problematising, group problem-solving, and the design process were supported through teacher and student questioning.

Next steps in STEM education research in Asia

While the three chapters may not represent a comprehensive review of STEM education research in Asian economies, the insights provided point to several directions for further research. In this section, I first summarise the authors' viewpoints of aspects that warrant further research. I then highlight additional STEM education research agendas that could be pursued in these three economies and other economies in similar existing states of research.

Mohd Saat and Mohd Fadzil (Chapter 10) called for greater attention on multidisciplinary STEM research in Malaysia, as multidisciplinarity better reflects real-life contexts as natural phenomena are not bound by a single STEM discipline. The authors also noted the need for more rigorous research in STEM education at various levels – including a wide spectrum of participants, subjects, and methodologies – to generate sound empirical evidence that would inform the progress of STEM education in Malaysia. Hong (Chapter 11) proposed three future directions for STE(A)M education in Korea. Firstly, basic education in science, mathematics, and informatics should be strengthened with an aim of increasing the current low levels of student interest and confidence. Secondly, a flexible learning environment for online and offline learning with ongoing updating of teaching materials and teachers' technological skillset is necessary to facilitate future-oriented education. Thirdly, the establishment of a sustainable ecosystem for science, mathematics, and informatics education through the concerted efforts of individuals, schools, and society, including partnerships with universities, companies, public institutions, STEM-related museums/centres, etc. All three proposed future directions point to necessary further research, fundamentally in their effectiveness in meeting the goals of STE(A)M education in Korea. Finally, Tan and Lau (Chapter 12) emphasised the need for greater understanding of interactions among learners and between teachers and learners, in order to identify and enact productive integrated STEM teaching strategies. The authors also called for space and time within STEM classrooms to facilitate learners in making sense of complex, persistent, and extended problems in STEM activities before proceeding with generating solutions.

Collectively, the three sets of authors attended to STEM education research needs in three different aspects corresponding to the focus of their chapters: Research methodology (Chapter 10), macro-level learning environment and ecosystem design (Chapter 11), and micro-level classroom interactions (Chapter 12). Here, I propose two additional research agendas for consideration.

Research methodologies for STEM practices

A review of 49 empirical research articles on assessment of student learning in integrated STEM education by Gao et al. (2020) indicates most assessments focused on students' monodisciplinary content knowledge and affect (i.e. attitudes, beliefs, motivation, and interest towards a STEM discipline) or transdisciplinary affect (i.e. interest in/attitude towards STEM disciplines or careers). While many reviewed research programmes aimed to improve students' interdisciplinary understanding or skills, their assessments did not match up. As a case in point, among the many studies that developed STEM learning activities based on engineering design, few (eight out of 38) assessed students' engagement in engineering design practices (Gao et al., 2020). Moreover, assessment of final products was the most common approach for assessing engineering practices. Unfortunately, such assessment misses out on the critical process of iterative refinement in engineering design (Gao et al., 2020). Acknowledging the challenges of assessing practices, Gao and colleagues recommended constructing a framework of STEM practices with a corresponding set of assessment methods. Examples of proposed STEM practices include engineering design, problem-solving, interdisciplinary reasoning and communication processes, and collaboration. Hence, complementing the call by Mohd Saat and Mohd Fadzil (Chapter 10) to widen the spectrum of participants, subjects, and methodologies, STEM education researchers should also consider the validity of their research design and analytical methods in view of the targeted goals of studied STEM activities or curricula.

Local agency in STEM education

One aspect that appears to be missing from the discourses in the three-book chapters is local agency in STEM education. In Chapter 10, Mohd Saat and Mohd Fadzil informed us of the limited research on STEM outreach programmes (e.g., STEM seminars, industry visits, competitions) in Malaysian-based STEM research (23 out of 89 articles) as majority of STEM research was based on researcher-initiated programmes. The finding calls for further understanding of the nature and extent of such STEM outreach programmes (e.g., the types of programmes and their prevalence); who initiated them and why, and the nature of participation (e.g., who participated and why; whether self-selected or teacher/school-selected/nominated). Additionally, while it is unlikely that once-off STEM outreach programmes would have significant impact on student outcomes such as improvement in students' STEM interest/attitude or development of disciplinary content knowledge or STEM practices, it would be of research value to compare the impacts of STEM outreach programmes versus researcher-initiated programmes in the abovementioned outcomes. In Chapter 11, Hong reported on various seemingly "top-down" approaches for supporting integrated STEAM education disseminated from the ministerial level to school level. What, might be examples of "ground-up" integrated STEAM programmes that meet local needs? These would include STEAM programmes not driven by the national education policies or national centres and instead, be initiated by researchers or local schools, other institutions, or organisations. What, then, would be the impact of top-down versus ground-up STEAM programmes on various student outcomes? Although not made explicit in Chapter 12, based on insider perspective, there exists a mix of top-down planning and ground-up initiatives of STEM programmes in Singapore. Thus, the same set of research questions are relevant to Singapore-based STEM research studies. Furthermore, studies could also focus on the extent of student agency in determining the pathways of their integrated STEM education, including choices and decisions students get to make within/across STEM programmes and activities. Indeed, student agency has been identified as a core concept associated with students learning how to learn in the Organisation for Economic Co-operation and Development [OECD]'s Future of Education and Skills 2030 project (OECD, 2019). Notably, OECD's concept note on student agency recognises the differences in interpretations of "agency" in different cultures. For example, Asian cultures tend to value agency in service of achieving collective harmony within communities rather than for achieving personal goals which are valued in Western cultures (OECD, 2019). Thus, research on local agencies in STEM education, be it demonstrated by local schools, other institutions or organisations, as well as students or teachers, need to consider the cultural context within which the research STEM education system is situated. Key considerations include: What does "agency" mean to different stakeholders of STEM education and how is it valuable or meaningful? What does local agency at various levels (e.g., individual students, class, teacher, school, community) look like and to what extents are they encouraged? Are the extents of local agencies adequate?

The tension of "doing School" versus "doing STEM"

Appealing as the abovementioned potential research agendas might sound, many STEM education researchers in Asia have probably come up against a key issue in redesigning learning environments to foster student participation in STEM education: To design integrative STEM curriculum or units of work that "improve or at least preserve students' examination results while concurrently fostering their creative and collaborative authentic problem-solving capacities" (Lee et al., 2019, p. 3). As noted by Lee and colleagues, this calls for "very powerful pedagogical design" (Lee et al., 2019, p. 3). In Asian economies, where high-stake standardised assessments serve as a tool for maintaining meritocracy, it appears that 21st century or future-oriented competencies (e.g., creative and critical thinking; collaborative problem-solving) targeted by STEM education cannot be developed at the expense of students' performance in high-stake standardised assessments. Herein lies the tension between "doing school" (Jiménez-Aleixandre et al., 2000) versus "doing STEM". Conventional school problems in high-stake standardised summative assessments often have a single correct answer, which is desirable for the purpose of standardised summative assessment. On the other hand, STEM problems are real-world problems at the individual, community, societal, national, or global level that could be solved using combinations of conceptual knowledge, procedural skills, and epistemic and social practices within and across individual STEM disciplines (Pleasants, 2020) i.e. ways of thinking, knowing, and doing in STEM. STEM problems involve an authentic audience with a persistent and complex problem waiting to be solved (Tan et al., 2019). Thus, STEM problems do not have a single correct solution, although some solutions would be better at addressing the problem than others. Given limited resources (including time), expecting students to excel in answering conventional high stake assessment questions i.e., "doing school" as well as solving STEM problems i.e., "doing STEM" would pull students, teachers, and parents in different directions.

A way out of this conundrum is to take a step back and ask ourselves, *what knowledge is of most worth in the Fourth Industrial Revolution?* (Zhao, 2018). The World Economic Forum (n.d.) describes the Fourth Industrial Revolution as "representing a fundamental change in the way we live, work, and relate to one another". As argued by Zhao (2018), if we acknowledge that education should be future-oriented i.e. to equip future members of societies with the knowledge and skills necessary in the new society, then it becomes apparent that the existing conventional education paradigm aimed at equipping all learners with the same knowledge and skillsets – relevant to the current or past society – to produce a homogenous workforce with similar abilities (which, I would add, is reinforced through high-stake standardised assessment) is inadequate for preparing the new generation for the future society. On the other hand, integrated STEM education – framed with the goal of inculcating students with 21st century or future-oriented competencies – seems a more viable approach for future-oriented education, with the caveat that it is envisioned as a model for a new education paradigm rather than a quick-fix solution to make students future-ready.

Integrated STEM education: A new paradigm of education?

To figure out how good the existing integrated STEM education model is (here, we consider what is typical across educational systems) as a candidate for the new education paradigm, we should first consider the alignment of their goals. Two worthy goals of future-oriented education include preparing students for jobs that cannot be done by machines (*including artificial intelligence*) and "help[ing] future generations to live peacefully and productively with others *and co-exist sustainably with the environment,* no matter where they are located" (Zhao, 2015, p. 26) (my addition in italics). As most integrated STEM education programmes target development of learners' future-oriented competencies, there appears to be broad alignment goals-wise. However, as critiqued by Pleasants (2020), many integrated STEM education programmes focused on STEM problems with strong emphasis on STEM disciplinary concepts and practices without attending to the social, cultural, political, or ethical (i.e., non-STEM) dimensions. Hence, for integrated STEM education to achieve the goals of future-oriented education, learners need to be presented with expanded STEM problems that incorporate non-STEM dimensions which exist in complex, real-world problems (Pleasants, 2020), including deliberations of the positive and negative impacts of STEM solutions on individuals and societies (Tan, 2020).

If we stop at changing the nature of STEM problems presented to learners, then we have merely applied a quick-fix solution. Future-oriented education needs a paradigm shift, a change in our worldview of the nature of education and what education looks like. As an example, Zhao (2015) proposed the idea of a standardised national or local curriculum, prevalent in many Asian education systems, would be unnecessary if we consider education as a personalised, student-centred experience. To begin envisioning what a new model of integrated STEM education and new paradigm of education might look like, we can consider the approaches of four successful Asian education systems – Hong Kong, Shanghai, Singapore, and South Korea – in shifting towards future-oriented education. According to Zhao (2015), these four systems have been working on integrating the new and the old, restructuring curricula, and balancing mandates and choices. While not originally intended for STEM education, these approaches seem equally relevant for future-oriented integrated STEM education. The approaches are outlined as follows. Interested readers should refer to Zhao (2015) for a detailed explanation.

Firstly, *pursue worthwhile outcomes.* As previously discussed, the knowledge most worth pursuing in preparation for the future is unlikely to be found in standardised high-stake assessments. If we can accept the reality of assessments that the most worthy outcomes may not have valid and reliable measures, then we should invest our efforts in outcomes that matter and come up with innovative, indirect (Zhao, 2015), or proxy measures. Secondly, *encourage ground-up initiatives and reduce top-down planning*. For new education models to emerge, experimentation at the local level should be encouraged. Various stakeholders including school leaders, teachers, students, parents, and communities should have the space to create new educational practices (c.f. proposed research agenda, local agency in STEM education). Thirdly, *grant meaningful autonomy to schools, teachers, and students*. This is essential for encouraging ground-up initiatives. According to Zhao (2015), the four successful Asian education systems have granted greater autonomy to schools and teachers to exercise their professional judgement over curriculum content and implementation. Students have also taken on greater responsibility for their own learning with reduced required examinations and courses. Finally, *build on strengths and respond to unique contexts*. For each Asian education system to be successful, it should learn from other countries and also recognise the uniqueness of its educational tradition and history. As Zhao (2015) rightly pointed out, successful examples within a system may provide the foundation for building a better future. Thus, each system should capitalise on its own strengths as the foundation for its STEM education.

In conclusion, using the existing integrated STEM education in your system as an initial model, I invite STEM education researchers to initiate conversations with other (STEM) education stakeholders around co-constructing a vision for a new education paradigm and to take actions to materialise this new paradigm in your respective education systems. As we approach the end of the first quarter of the 21st century, we have to ask ourselves to what extent have our learners acquired 21st-century competencies and how ready are they for the future? The answers indicate the pace at which we have to work towards materialising a future-oriented integrated STEM education within our system. Dauting as the task may seem, as Nelson Mandela once said, "it is only impossible until it's done".

References

Gao, X., Li, P., Shen, J., & Sun, H. (2020). Reviewing assessment of student learning in interdisciplinary STEM education. *International Journal of STEM Education*, 7(24). https://doi.org/10.1186/s40594-020-00225-4. https://doi.org/10.1186/s40594-020-00225-4

Jiménez-Aleixandre, M. P., Bugallo Rodríguez, A., & Duschl, R. A. (2000). "Doing the lesson" or "doing science": Argument in high school genetics. *Science Education*, 84(6), 757–792. https://doi.org/10.1002/1098-237X(200011)84:6<757::AID-SCE5>3.0.CO;2-F

Lee, M. H., Chai, C. S., & Hong, H. Y. (2019). STEM education in Asia Pacific: Challenges and development. *Asia-Pacific Education Researcher*, 28(1), 1–4. https://doi.org/10.1007/s40299-018-0424-z

National Science Foundation. (n.d.). *Convergence research at NSF.* Retrieved May 10, 2021, from https://www.nsf.gov/od/oia/convergence/index.jsp

Organisation for Economic Co-operation and Development. (2019). *Conceptual learning framework: Student agency for 2030.* https://www.oecd.org/education/2030-project/teaching-and-learning/learning/student-agency/Student_Agency_for_2030_concept_note.pdf

Pleasants, J. (2020). Inquiring into the nature of STEM problems: Implications for pre-college education. *Science and Education, 29*(4), 831–855. https://doi.org/10.1007/s11191-020-00135-5

Takeuchi, M. A., Sengupta, P., Shanahan, M. C., Adams, J. D., & Hachem, M. (2020). Transdisciplinarity in STEM education: A critical review. *Studies in Science Education, 56*(2), 213–253. https://doi.org/10.1080/03057267.2020.1755802

Tan, A. L., Teo, T. W., Choy, B. H., & Ong, Y. S. (2019). The S-T-E-M Quartet. *Innovation and Education, 1*(1), 3. https://doi.org/10.1186/s42862-019-0005-x

Tan, M. (2020). STEM as opportunity to get TSLN right: Science education for economically productive creativity. *Asia Pacific Journal of Education, 40*(4), 485–500. https://doi.org/10.1080/02188791.2020.1838882

World Economic Forum. (n.d.). *Fourth Industrial Revolution.*

Zhao, Y. (2015). Lessons that matter: What should we learn from Asia's school systems? In *Mitchell Institute.* http://www.mitchellinstitute.org.au/reports/lessons-that-matter-what-should-we-learn-from-asias-school-systems

Zhao, Y. (2018). Shifting the education paradigm: Why international borrowing is no longer sufficient for improving education in China. *ECNU Review of Education, 1*(1), 76–106. https://doi.org/10.30926/ecnuroe2018010105

10 STEM education research in Malaysia

Rohaida Mohd Saat and Hidayah Mohd Fadzil

Introduction

The trend in the international publications related to STEM education proves how scholarship in science, technology, engineering, and mathematics (STEM) education has grown tremendously. With global awareness of the rising importance of STEM education, educators responded to this call by publishing their scholarly work across a variety of publication platforms (Li, 2018). Analysis by Li et al. (2020) indicates that STEM education research has been increasingly recognised as an important research field, and numerous studies have been published, particularly over the last decade. The rapid increase in the number of scholarly publications on STEM education in recent years has led to the need to examine trends in STEM education research in Malaysia. By examining the trend, it might provide some indicators to enhance STEM education in Malaysia further. In educational research, it is common to conduct literature reviews to examine the status and developments in particular disciplines. By knowing and recognising the status and developments in STEM education, it can be further developed and properly supported by the relevant stakeholders. This principle also applies to STEM education.

Research on STEM education is an area of broad variety and uncertain criteria. Unlike discipline-based education research, STEM education is not a well-defined field (Li et al., 2020). The disparity of STEM education can be seen in many different definitions of STEM education (Brown, 2012; Jayarajah et al., 2014). There are many reviews conducted in STEM-related fields over the years, and they are usually discipline-based, for example, in science education (e.g., Erduran et al., 2015; Lin et al., 2018; Minner et al., 2010), mathematics education (e.g., Bray & Tangney, 2017; Kilpatrick, 1992; Sokolowski et al., 2015), technology and engineering education (e.g., Borrego et al., 2015; Xu et al., 2019), and also in STEM education (e.g., Gao et al., 2020; Gül & Taşar, 2020; Jayarajah et al., 2014; Li et al., 2020). Therefore, undertaking an analysis of literature in STEM education research involves careful consideration and specifically defined scope. This is to resolve the difficulty inevitably associated with STEM education. In this chapter, we

DOI: 10.4324/9781003099888-10

define STEM education as a discipline by itself. Based on this definition, we systematically examined journal publications in STEM education research to take a closer look at the development of this discipline in the context of Malaysia.

Recent studies concerning school science and the desire to choose science-related careers among secondary school students is troubling, as students reject careers related to science as a potential career choice (Fadzil et al., 2019; Kudenko & Gras-Velázquez, 2016; van Griethuijsen et al., 2015; Zhou et al., 2019). According to Gilbert and Justi (2016), evidence of the lack of participation of students in science is used to support widespread discontent with the rate of achievement of students in international studies such as TIMSS (Trends in International Mathematics and Science Study) and PISA (International Student Evaluation Program). This has caused concern among policymakers about the science literacy of their populations, as well as the scientific and technical workforce of their country (van Griethuijsen et al., 2015).

In the case of Malaysia, the decrease in the number of students in the science stream is disturbing, and this may lead to a severe human capital shortage in the STEM field if the number of students who enrol in STEM courses does not meet its expected standards, which is 270,000 per year or 60% of the annual national cohort (Fadzil et al., 2019). Jayarajah et al. (2014) highlighted that Malaysia consistently registers lower numbers of interested citizens in pursuing science and technology as their career. In addition, performance of Malaysian science students has been disappointing in international benchmarking assessments such as TIMSS and PISA. Such performance will provide some indication of the students' future achievement in STEM-related courses at higher levels. Thus, this will affect the future of the science and technology labour force of Malaysia. Both factors, namely low number of students enrolling in the science stream and low performance, will impact the future of science or STEM education in Malaysia.

STEM-related subjects need to be taught in a more enriching and interesting way to keep curiosity going. Malaysian science education aims to make science more attractive for students and to indirectly invite more students to pursue their studies in STEM-related fields to realise Malaysia's ambition of becoming an industrialised country (Fadzil et al., 2019). In line with that, this chapter focuses on the research in STEM education in Malaysia over a period of ten years and the possibilities for future development of STEM education.

Therefore, this chapter will present the status and trends in Malaysia STEM education research from 2010 to 2020 based on journal publications. The trend of STEM education research in Malaysia is presented in four main areas which are: (1) The levels of education where the research was conducted; (2) Types of participants or subjects involved in the research, (3) Types of STEM activities, and (4) STEM research area.

Method of the literature review

Framework for the literature review

The review followed a structured and systematic approach. Conducting a literature review involves an explicit research approach that uses literature as evidence instead of observations, interviews, or questionnaires. This review follows the framework by vom Brocke et al. (2009), which consists of four phases; (1) Definition of review scope, (2) Conceptualisation of the topic, (3) Literature search, and (4) Literature analysis (Figure 10.1).

Phase 1: Definition of review scope

According to Vom Brocke et al. (2009), the study's review scope should centre on the research outcomes and theories of the literature. The scope of the formal literature review is very extensive with the goal of covering most of the literature within the specified scope, i.e., STEM education research in Malaysia within the time frame of ten years.

Phase 2: Conceptualisation of the topic

In this phase, we need to clarify the scope and conceptualise the topic. The present literature review intends to present an assessment of the STEM research that has emerged as a result of STEM education reform in Malaysia. To keep up with the STEM education development globally, the Malaysian government launched the Malaysia Education Blueprint (2013–2025), which aims to increase students' and teachers' interest, attitude, and motivation in STEM education, as well as career awareness in STEM. Despite the fact that STEM education has long been considered in the United States, the definition and idea of STEM is still relatively new in Malaysia (Bahrum et al., 2017). For this reason, the outset of the current study is the identification of key terms and topics from the Malaysia study that could be used in further search processes. The key terms include "Malaysia STEM education", "Malaysia STEM education research". We also include key terms in Malay Language, such as *pendidikan STEM di Malaysia* (Malaysian STEM education) and *kajian pendidikan STEM di Malaysia* (Malaysian STEM education research).

We used the acronym STEM or written as the phrase "science, technology, engineering, and mathematics" as a term in our search of publication titles. To

Figure 10.1 Four phases framework for analysis.

identify and select articles for review, we searched all items in the selected databases and chose only those articles that the author(s) self-identified with the acronym STEM (or written as the phrase "science, technology, engineering, and mathematics") in the title and/or abstract. The search was limited from the year 2010 until 2020. The ten-year timeframe was chosen to ensure that the findings of this study are still relevant, not out of date and are appropriate for use as a guide by stakeholders in the field of science education, as suggested by Winarno et al. (2020).

Phase 3: Literature search

In this phase, the published research papers were searched, evaluated, and selected. To retrieve the published articles involving STEM education in Malaysia, major databases in STEM Education were scanned, i.e., MyCite (Malaysian Citation Index), Scopus, Web of Science (WoS), and ERIC (Education Resources Information Centre) (Li et al., 2020). MyCite or Malaysian Citation Index is a citation system initiated by the Ministry of Education (MOE) Malaysia in 2011. MyCite provides access to bibliographic as well as full-text contents of scholarly journals published in Malaysia in the fields of Science, Technology, Medicine, Social Science, and the Humanities.

The next step, as shown in Table 10.1, involved the collection of the related contributions from the search result with 292 entries in the four databases. Papers that were indexed in multiple databases, were placed in the most reputable databases, in the following order: Web of Science, Scopus, ERIC, and MyCite. For example, if the paper was indexed in MyCite as well as in ERIC, the paper was counted in the ERIC database.

Phase 4: Literature analysis

This phase is the analysis phase, where the selected literature was analysed. This included first reviewing the title and the keywords, and the abstract if needed. All the authors scrutinised a total of 292 contributions via a collaborative process. We excluded publications in letters to editors, editorials and proceedings papers and this process resulted in the final selection of 203 contributions. Based on this criteria, the number of publication gathered articles (consisting of "included" and "maybe included") totalled 111 articles. The authors further examined the selected articles to eliminate the "maybe included" contributions and selected a net list of 89 contributions (refer to Table 10.1). Based on the papers' title, abstract, areas of study, samples, methodology, and findings, this chapter was able to provide empirical results, including the temporal distribution of STEM education in Malaysia, the levels of education where the research was conducted, types of participants involved in the research, types of STEM activities, and research area analysis.

In ensuring the validity of the result, Bryman (2008) suggested that each of the authors separately evaluated the result, thereby performing triangulation

Table 10.1 Databases and contributions of published articles

Keywords used	Type of articles	Databases			
		Web of Science (WoS)	Scopus	ERIC	MyCite
Malaysia STEM education; Malaysia STEM education research; *pendidikan STEM di Malaysia* (Malaysia STEM education); *kajian pendidikan STEM di Malaysia* (Malaysia STEM education research).	All types of articles	78 [Articles = 45 Proceedings = 29 Review = 2 Early Access = 1 Editorial material = 1 Letter = 1]	118 [Articles = 70 Proceedings = 35 Review = 5 Book chapter = 4 Editorial = 2 Letter = 1 Note = 1]	45 [Articles = 43 Review = 2]	51 [Articles = 45 Proceedings = 2 Review = 4]
	Articles only	45	70	43	45
	Education-related articles (included & maybe-included)	26	36 (46 before excluding redundant articles)	17 (37 before excluding redundant articles)	32 (39 before excluding redundant articles)
	Further analysis-related articles	22	31	12	24

(i.e., multiple investigators). Moreover, according to Singh (2017), this step of independent data abstraction can minimise the risk of errors during the literature review. This study adopted Bryman's (2008) suggestion. The two researchers worked independently and moderation was done at various phases of the study. Any disagreements or discrepancies were resolved through discussion.

Trend of STEM education research in Malaysia

The articles reviewed in this study were published from 2010 to 2020. Figure 10.2 shows the number of publications per year during the studied period. The total number of articles identified was 89, across a period of 10 years.

As Figure 10.2 shows, the number of publications increases each year beginning in 2016 (n = 7). A very limited number of articles concerning STEM education were published between 2010 and 2013 (only 1 article was published in 2012). In 2014, five articles were published, and all of them were indexed in the Web of Science database. However, the numbers decline to only two articles in 2015. The result shows that research in STEM education has grown significantly in 2016 (n = 7), 2017 (n = 8), 2018 (n = 20), and 2019 (n = 26).

The noticeable jump from 2017 to 2019 of STEM education publications also suggests that STEM education research has gained the recognition of

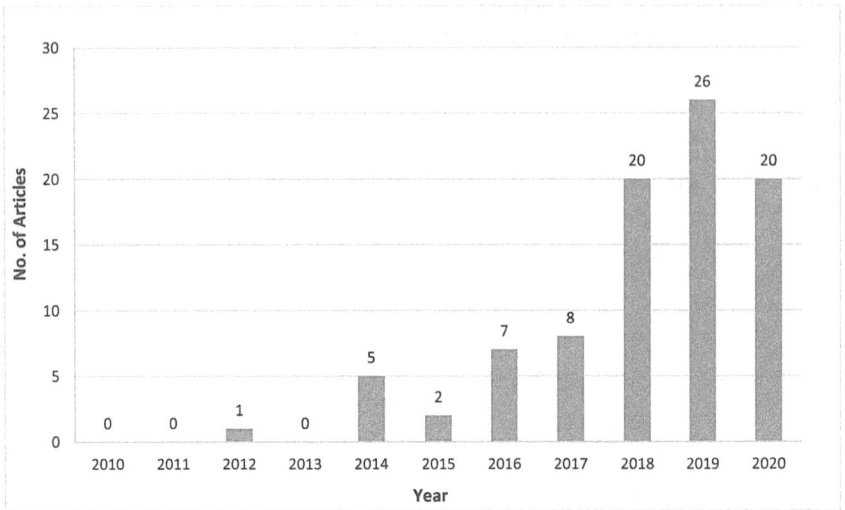

Figure 10.2 Distribution of reviewed articles based on year (2010–2020).

researchers in Malaysia. The highest frequency for the reviewed articles is in the year 2019 which contributes a total of 29% to the overall distribution of the reviewed articles.

Based on the total number of published articles (n = 89), 35% of them are from the Scopus database, 27% from MyCite, 25% from WoS, and 13% from the ERIC database (Figure 10.3). Collectively, the results suggest that many scholars tend to include the acronym STEM in their publications' titles to highlight

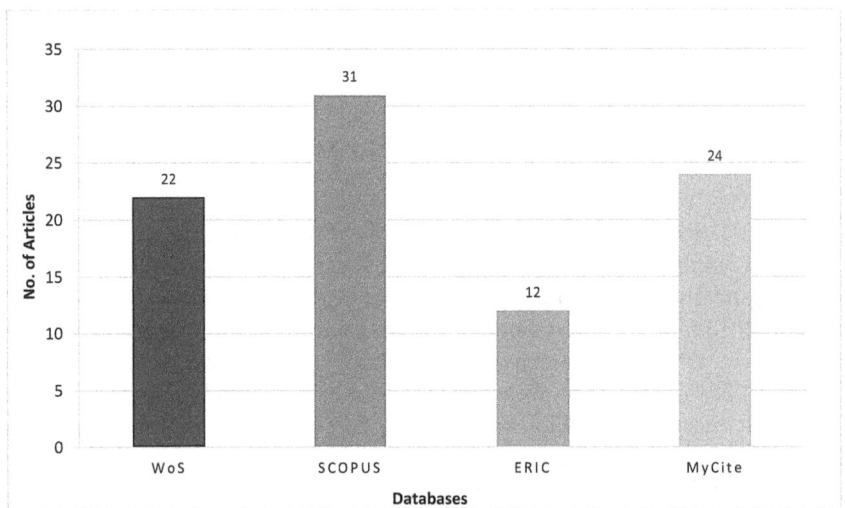

Figure 10.3 Number of total articles in selected databases (2010–2020).

their research in or about STEM education. It is noticeable that the use of the acronym STEM is growing at a faster rate in the Malaysian context. This is in comparison to previous reviews, for example, Jayarajah et al. (2014), who reviewed articles for over a period of 14 years (1999 to 2013) and found only 57 articles related to STEM education research in Malaysia. They highlighted a lack of research solely on STEM education and the acronym "STEM" was not included in their title or/and abstract. Thus they reviewed published articles in all the four individual STEM disciplines using key search terms such as "science and mathematics", "technology", "engineering", "education", and "field" (Jayarajah et al., 2014). The current review also shows that only one STEM education publication has been found from 2010 to 2013. This is possible as Malaysian researchers did not publish under the keyword "STEM". However, the recent trend shows that Malaysian researchers have started to include STEM in their title or/and abstract and indicates that various STEM education research has received their own recognition in the education field (Fan & Yu, 2015; Li et al., 2020; Winarno et al., 2020).

In addition, the increase in research in STEM education is likely due to the emphasis on STEM education via various policies and initiatives. For example, one of the policies is the National STEM Action Plan (2018–2025), where its main agenda is to advance Malaysia towards a more competitive and competent nation based on the strong fundamental of science, technology, and innovation. Another example is the rollout of the Malaysian Education Blueprint (2013–2025) in late 2012, where STEM Education was included in the first shift of the Blueprint to transform the education system. MOE Malaysia and other institutions have also established centres dedicated to STEM Education, such as the National STEM Centre. The establishment of such centres has created awareness of the importance of STEM education.

The rigour of STEM education research in the past ten years is evidenced by the increase in the number of refereed journals that focused on STEM education. A simple search from the four databases used in this literature review has generated about eight journals dedicated to STEM education, such as the International Journal of STEM Education, Journal for STEM Education Research, and European Journal of STEM Education. All these phenomena have probably contributed to an increased number of publications in STEM education among Malaysian researchers.

Classifications of STEM education research

The classification of STEM education research in Malaysia is presented in four main areas, which are:

1 The levels of education where the research was conducted.
2 Types of participant or subject involved in the research.
3 Types of STEM activities.
4 STEM research area.

Table 10.2 Level of research conducted in STEM education in Malaysia

Level	Number of studies
Pre-school	2
Primary school	10
Primary and secondary school	10
Secondary school	54
Pre-university/diploma	7
Tertiary	6
Total	**89**

The levels of education

STEM education cuts across all levels. In this context, the levels of education refer to the grade level where the STEM education research was conducted. There are six major groups of research levels that were emphasised in the published articles from Malaysia. These groups of participants vary from pre-school, primary school, secondary school, primary and secondary school, pre-university and diploma, and up to tertiary level. Table 10.2 describes the research level involved and the number of studies for each level.

The results suggest that the research community had a broad interest in both teaching and learning in pre-school to tertiary level STEM education. It is also observed that STEM education researchers in Malaysia focus more on the secondary school level, with 61% (n = 54) of the total number of publications analysed in this study.

The result is coherent with Malaysian STEM policy, which strives to achieve the target of having 60% of upper secondary students pursuing STEM-related academic stream. Based on previous studies (e.g., Fadzil et al., 2019; Shahali et al., 2017), more students chose to be in the science or STEM-related academic stream for their upper secondary education and chose a STEM-related career.

The research trend demonstrated in Table 10.3 also discloses the extent of importance given to primary school level as more articles has been published using primary school students (primary, 11%; primary and secondary, 11%), compared to previous years, as suggested by Jayarajah et al. (2014). In the review by Jayarajah et al. (2014) on STEM education research from 1999–2013, only four research articles involved primary school students. The small number of studies at the pre-school level and primary school correspond with the current situation of science education studies concerning children in Malaysia. Previous studies (e.g., DeJarnette, 2018; Jayarajah et al., 2014) highlight that few opportunities exist for children and their teachers in STEM education at early education.

According to Talafian et al. (2019), children do not see themselves in STEM roles or establish STEM identities by the end of primary school. It is important to capture students' interest in STEM content at an early age to ensure that students are on track through primary and secondary school as adequate

Table 10.3 Types of participants or subjects involved in the studies

Participants		WoS	Scopus	ERIC	MyCite	Total
Students	Pre-school	-	-	-	-	-
(School)	Primary	3	-	-	3	6
	Secondary	7	9	7	12	35
Post-secondary students	Pre-university	1	3	1	1	6
Graduates	Undergraduate	-	3	-	-	3
(University/ college)	Postgraduate	1	-	-	-	1
Practitioners	Teacher/lecturer	-	4	-	2	6
Others:	More than one type of participants	10	12	4	6	32
Total		22	31	12	24	89

preparation can be accomplished (DeJarnette, 2018; Hachey, 2020). This also demonstrates the need to concentrate on the participation of young children in STEM education as a way of promoting their early development of STEM identity. More articles on STEM education should be published in the context of Malaysian pre-school and primary school levels.

Types of participants or subjects involved in the research

Although the previous area focused on the level of education that STEM education research was conducted, this subsection provides a different perspective as the participants involved did not necessarily correspond to the level of education. Table 10.3 describes the types of participants involved, and the number of participants gathered in each STEM discipline. There are five major groups of participants that were emphasised in the published research articles in STEM education. These groups of participants are school level students, post-secondary school students (i.e. pre-university), university graduates, practitioners (i.e. teachers and lecturers), and others studies with more than one type of participants, for example, a mixture of primary and secondary school students, or postgraduate students with school teachers as participants of the study.

It is observed that STEM education researchers in Malaysia have emphasised school students as participants in their studies, with a total of 46% of the published articles. The research trend discloses that no study has given importance to pre-school children as the main participants. However, there are studies that focused on pre-school STEM teachers together with the children as participants instead. From the trend, it is perceived that researchers in Malaysia prefer to employ more mature participants in their research, such as from secondary school, as suggested by Jayarajah et al. (2014).

On the other hand, only a handful of studies have used practitioners, i.e. teachers and lecturers, as the main participants, although they are more mature participants. Previous studies (e.g., Ramli & Talib, 2017; Siew et al., 2015)

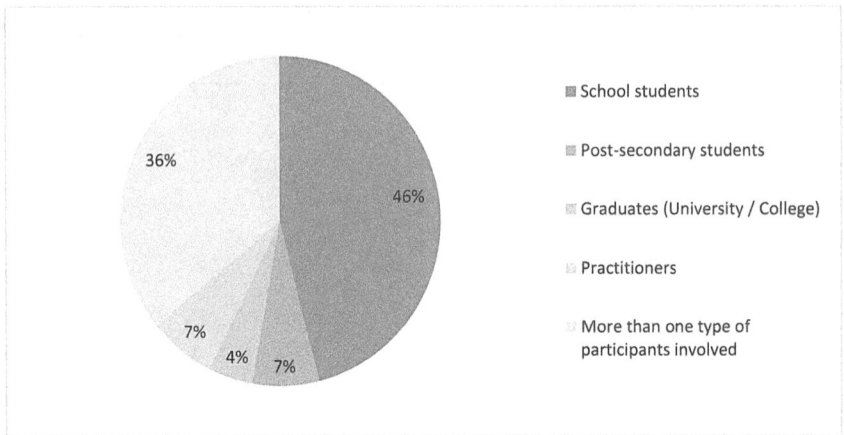

Figure 10.4 Types of participants in Malaysia STEM studies by percentage (2010–2020).

reported that most science teachers are still struggling to make connections across the STEM disciplines. The lack of understanding of STEM may lead to a sense of incompetence among science teachers. In the case of early childhood teachers, recent studies (e.g., DeJarnette, 2018; Tao, 2019) also reported that the teachers have low self-efficacy for teaching STEM, resulting in limited time devoted to teaching STEM to the children. Therefore, further studies should be carried out among Malaysian practitioners at every level to establish a positive effect on the future of Malaysia's STEM education. For instance, appropriate intervention can be done with practitioners at different levels. It is necessary to involve teachers in extensive professional development programmes, which include the use of successful strategies to reach the ultimate goal of teacher's professionalism in STEM education, as proposed by Aldahmash et al. (2019). For example, primary school teachers in Malaysia probably need more training on STEM content because they are previously trained in general science content. For secondary school, the teachers need more training on STEM pedagogy and the integration of STEM elements in teaching and learning science.

It is also observed that Malaysian researchers tend to include more than one type of participant in their studies (36%) (See Figure 10.4). The combinations of participant types occur when more than one type is involved, such as teachers and primary school students, teachers, and secondary school students and teachers, scientists, and students as participants. This may be due to the way the researchers triangulate their data, as data collection from multiple participants can be considered as a type of triangulation, where each participant brings their own perspectives and worldview, offering new dimensions and the possibility of experiencing commonalities (Saat & Fadzil, 2019). However, this assumption needs to be validated by further reviewing the studies in this category to determine if data triangulation was indeed carried out with multiple types of participants.

Types of activities in STEM education studies (2010-2020)

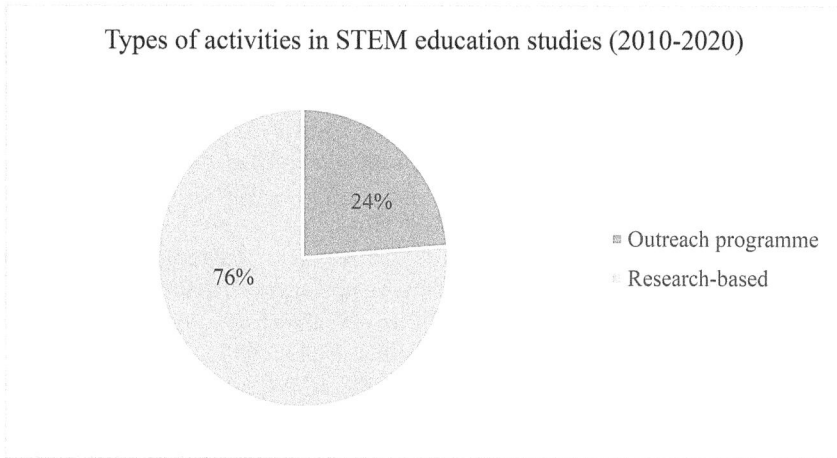

Figure 10.5 Types of activities in Malaysia's STEM education studies (2010–2020).

Types of STEM activities

The types of STEM activities can be divided into two main categories, which are research-based activities and outreach programmes, as summarised in Figure 10.5.

In the context of this review, research-based activities are activities related to studies that use intervention and these include studies with theoretical and empirical insights (Wilson, 2014). They also include activities with an investigative element of research and have evidence for the particular concept or model used. Based on the review, 76% (n = 66) of the published articles used activities that have the element of research. On the other hand, outreach programmes involve events such as STEM seminars, industry visits, competitions, and activities geared towards making STEM subjects captivating. Examples of practitioners involved in outreach programmes include individuals from professional institutions, non-profit organisations, or higher educational institutes where they provide crucial opportunities for students to engage in real-world STEM challenges (Aslam et al., 2018; Vennix et al., 2018) and aim to motivate students to pursue a career in STEM by connecting STEM with school science. These activities are often added to the regular curriculum (Vennix et al., 2018). Based on the total number of 89 publications, 23 articles were categorised as involving outreach programmes (24%). This might indicate that outreach activities in STEM education receive a good response within the STEM research community, as STEM outreach programmes generally have positive impacts on participants' understanding of STEM and their attitude toward STEM disciplines, as suggested by a number of studies concerning STEM outreach (e.g., DeWilde et al., 2019, Zhou, 2020).

Research area or content scope

Figure 10.6 represents the number of studies categorised by the number of STEM areas involved in the study (i.e., science, technology, engineering, and mathematics) in terms of the research area or content scope being focused on. Based on the review, not all of the publications that used the STEM acronym include all of the four STEM areas. Thus, for each study, the number of STEM areas involved was further explored.

There are five categories of the research area of STEM education from 2010 to 2020. The categorisation of STEM content area was based on Li et al. (2020). Category 1 includes studies in only one area of STEM, for example, the activities that are related to one science discipline, either biology, physics, or chemistry. Category 2 involves studies with two areas of STEM, Category 3 includes three areas of STEM, Category 4 comprises studies from four areas of STEM, and Category 5 comprises studies with STEM and non-STEM research area. Studies involving only one area of STEM are the most published with 36 publications (41%), followed by studies involving two areas (25%), then studies involving four areas of STEM (21%), studies involving three areas of STEM (11%), and finally studies that involve both STEM and non-STEM (2%) subjects.

The highest number of studies in Category 1 likely reflects Malaysia researchers' interest in conducting individual disciplinary based studies. The lowest number of studies are those published in Category 5, which further affirms this claim. This shows that researchers in Malaysia might still find it difficult to conduct multidisciplinary and interdisciplinary scholarship in STEM education. This is in contrast to recent studies by Li and Schoenfeld (2019) and Li et al. (2020). Li and Schoenfeld (2019) noted that there is a growing interest in interdisciplinary and integrated STEM in international STEM education studies.

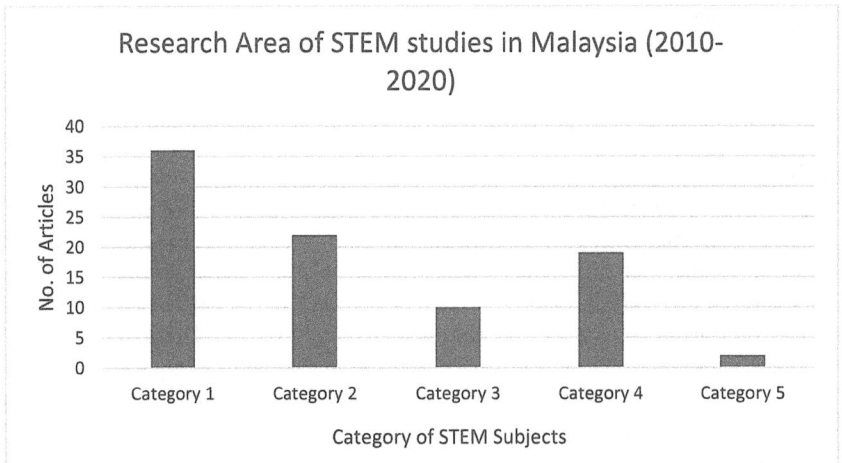

Figure 10.6 Publication distribution in terms of research area or content scope.

Conclusion

STEM education has become more known in the field of education as STEM-related fields have impacted other areas such as economics and politics. Despite various efforts to promote STEM education in Malaysia, STEM education still faces issues such as being an unpopular field of study. Such issues will affect the nation's agenda in addressing the challenges and demands of a science and technology-driven economy. This chapter introduces readers to the status and trend of research in STEM Education in Malaysia; according to the levels of education where the research was conducted, types of participants or subjects, types of STEM activities, and STEM area of research. The findings imply that STEM research in Malaysia needs to be more multidisciplinary in nature in line with the nature of STEM as a discipline. Malaysian STEM researchers need to have a mind shift to migrate to more multidisciplinary STEM research. Multidisciplinary STEM reflects the real-life setting as nature or natural phenomena involves more than one discipline of STEM.

STEM research in Malaysia should also equally focus on all levels of education and not focus disproportionately on secondary schools. More rigorous research in STEM education at various levels involving a wide spectrum of participants or subjects and a variety of methodologies can be employed to provide sound empirical evidence on how a nation should move forward, particularly in addressing the issues related to STEM education.

References

Aldahmash, A. H., Alshamrani, S. M., Alshaya, F. S., & Alsarrani, N. A. (2019). Research Trends in In-Service Science Teacher Professional Development from 2012 to 2016. *International Journal of Instruction, 12*(2), 163–178.

Aslam, F., Adefila, A., & Bagiya, Y. (2018). Stem Outreach Activities: An Approach to Teachers' Professional Development. *Journal of Education for Teaching, 44*(1), 58–70. doi: https://doi.org/10.1080/02607476.2018.1422618.

Bahrum, S., Wahid, N., & Ibrahim, N. (2017). Integration of STEM Education in Malaysia and Why to Steam. *International Journal of Academic Research in Business and Social Sciences, 7*(6), 645–654. doi: https://doi.org/10.6007/ijarbss/v7-i6/3027.

Borrego, M., Foster, M. J., & Froyd, J. E. (2015). What is the State of the Art of Systematic Review in Engineering Education? *Journal of Engineering Education, 104*(2), 212–242. doi: https://doi.org/10.1002/jee.20069.

Bray, A., & Tangney, B. (2017). Technology Usage in Mathematics Education Research – A Systematic Review of Recent Trends. *Computers & Education, 114*, 255–273. doi: https://doi.org/10.1016/j.compedu.2017.07.004.

Brown, J. (2012). The Current Status of STEM Education Research. *Journal of STEM Education: Innovations and Research, 13*(5), 7–11. https://www.jstem.org/jstem/index.php/JSTEM/article/view/1652.

Bryman, A. (2008). 6 Why do Researchers Integrate/Combine/Mesh/Blend/Mix/Merge/Fuse Quantitative and Qualitative Research? *Advances in Mixed Methods Research, 21(8)*, 87–100. SAGE Publications Ltd.

DeJarnette, N. K. (2018). Implementing STEAM in the Early Childhood Classroom. *European Journal of STEM Education*, *3*(3), 1–9. doi: https://doi.org/10.20897/ejsteme/3878.

DeWilde, J. F., Rangnekar, E. P., Ting, J. M., Franek, J. E., Bates, F. S., Hillmyer, M. A., & Blank, D. A. (2019). Evaluating Large-Scale STEM Outreach Efficacy with a Consistent Theme: Thermodynamics for Elementary School Students. *ACS Omega*, *4*(2), 2661–2668. doi: https://doi.org/10.1021/acsomega.8b03156.

Erduran, S., Ozdem, Y., & Park. J. Y. (2015). Research Trends on Argumentation in Science Education: A Journal Content Analysis from 1998–2014. *International Journal of STEM Education*, *2*(5), 1–12. doi: https://doi.org/10.1186/s40594-015-0020-1.

Fadzil, H. M., Saat, R. M., Awang, K., & Adli, D. S. H. (2019). Students' Perception of Learning STEM-Related Subjects through Scientist-Teacher-Student Partnership (STSP). *Journal of Baltic Science Education*, *18*(4), 537–548. doi: https://doi.org/10.33225/jbse/19.18.537.

Fan, S. C., & Yu, K. C. (2015). How an Integrative STEM Curriculum Can Benefit Students in Engineering Design Practices. *International Journal of Technology and Design Education*, *27*(1), 107–129. doi: https://doi.org/10.1007/s10798-015-9328-x.

Gao, F., Li, L., & Sun, Y. (2020). A Systematic Review of Mobile Game-Based Learning in STEM Education. *Educational Technology Research and Development*, *68*(4), (June), 1791–1827. doi: https://doi.org/10.1007/s11423-020-09787-0.

Gilbert, J. K., & Justi, R. (2016). *Modelling-Based Teaching in Science Education*. Cham: Springer International Publishing.

Gül, K. S., & Taşar, M. F. (2020). A Review of Researches on STEM in Preservice Teacher Education. *İlköğretim Online – Elementary Education Online*, *19*(2), 515–539. doi: https://doi.org/10.17051/ilkonline.2020.689682.

Hachey, A. C. (2020). Success for All: Fostering Early Childhood STEM Identity. *Journal of Research in Innovative Teaching & Learning*, *13*(1), 135–139. doi: https://doi.org/10.1108/jrit-01-2020-0001.

Jayarajah, K., Saat, R. M., & Rauf, R. A. A. (2014). A Review of Science, Technology, Engineering & Mathematics (STEM) Education Research from 1999–2013: A Malaysian Perspective. *EURASIA Journal of Mathematics, Science & Technology Education*, *10*(3), 155–163. doi: https://doi.org/10.12973/eurasia.2014.1072a.

Kilpatrick, J. (1992). A History of Research in Mathematics Education. In Grouws D. (Ed.), *Handbook of Research on Mathematics Teaching and Learning* (pp. 3–38). New York: Macmillan.

Kudenko, I., & Gras-Velázquez, A. (2016). The Future of European STEM Workforce: What Secondary School Pupils of Europe Think about STEM Industry and Careers. *Insights from Research in Science Teaching and Learning* (pp. 223–236). Cham, Switzerland: Springer.

Li, Y. (2018). Journal for STEM Education Research – Promoting the Development of Interdisciplinary Research in STEM Education. *Journal for STEM Education Research*, *1*, 1–6. doi: https://doi.org/10.1007/s41979-018-0009-z.

Li, Y., & Schoenfeld, A. H. (2019). Problematizing Teaching and Learning Mathematics as "Given" in STEM Education. *International Journal of STEM Education*, 6, Article 44. doi: https://doi.org/10.1186/s40594-019-0197-9.

Li, Y., Wang, K., Xiao, Y., & Froyd, J. E. (2020). Research and Trends in STEM Education: A Systematic Review of Journal Publications. *International Journal of STEM Education*, 7, Article 11. doi: https://doi.org/10.1186/s40594-020-00207-6.

Lin, T. J., Potvin, P., & Tsai, C. C. (2018). Research Trends in Science Education from 2013 to 2017: A Systematic Content Analysis of Publications in Selected Journals. *International Journal of Science Education*, *41*(3), 367–387. doi: https://doi.org/10.1080/09500693.2018.1550274.

Minner, D. D., Levy, A. J., & Century, J. (2010). Inquiry-Based Science Instruction – What Is It and Does It Matter? Results from a Research Synthesis Years 1984 to 2002. *Journal of Research in Science Teaching*, *47*(4), 474–496. doi: https://doi.org/10.1002/tea.20347.

Ramli, N. F., & Talib, O. (2017). Can Education Institution Implement STEM? From Malaysian Teachers' View. *International Journal of Academic Research in Business and Social Sciences*, *7*(3), 721–732. doi: https://doi.org/10.6007/IJARBSS/v7-i3/2772.

Saat, R. M., & Fadzil. H. M. (2019). Methodological Dilemma in Qualitative Research in Education. *The Malaysian Journal of Qualitative Research*, *5*(2), 41–46. https://www.qramalaysia.org/journal-vol5-no2.

Shahali, E. H. M., Ismail, I., & Halim, L. (2017). STEM Education in Malaysia: Policy, Trajectories and Initiatives. *Asian Research Policy Science and Technology Trends*, *8*(2), 122–133.

Siew, N. M., Amir, N., & Chong. C. L. (2015). The Perceptions of Pre-Service and In-Service Teachers Regarding a Project-Based STEM Approach to Teaching Science. *SpringerPlus*, *4*(8), 1–20. doi: https://doi.org/10.1186/2193-1801-4-8.

Singh, S. (2017). How to Conduct and Interpret Systematic Reviews and Meta-Analyses. *Clinical and Translational Gastroenterology*, *8*(5), e93. doi: https://doi.org/10.1038/ctg.2017.20.

Sokolowski, A., Li, Y., & Willson, V. (2015). The Effects of Using Exploratory Computerized Environments in Grades 1 To 8 Mathematics: A Meta-Analysis of Research. *International Journal of STEM Education*, *2*(8), 1–17. doi: https://doi.org/10.1186/s40594-015-0022-z.

Talafian, H., Moy, M. K., Woodard, M. A., & Foster, A. N. (2019). STEM Identity Exploration through an Immersive Learning Environment. *Journal for STEM Education Research*, *2*, 105–127. doi: https://doi.org/10.1007/s41979-019-00018-7.

Tao, Y. (2019). Kindergarten Teachers' Attitudes toward and Confidence for Integrated STEM Education. *Journal for STEM Education Research*, *2*, 154–71. doi: https://doi.org/10.1007/s41979-019-00017-8.

Van Griethuijsen, R. A. L. F., Van Eijck, M. W., Haste, H., Den Brok, P. J., Skinner, N. C., Mansour, N., Gencer, A. S., & BouJaoude, S. (2015). Global Patterns in Students' Views of Science and Interest in Science. *Research in Science Education*, *45*(4), 581–603.

Vom Brocke, J., Simons, A., Niehaves, B., Niehaves, B., Reimer, K., Plattfaut, R., & Cleven, A. (2009). Reconstructing the giant: On the importance of rigour indocumenting the literature search process. *European Conference on Information Systems (ECIS)*. https://aisel.aisnet.org/cgi/viewcontent.cgi?article=1145&context=ecis2009

Vennix, J., Den Brok, P., & Taconis, R. (2018). Do Outreach Activities in Secondary STEM Education Motivate Students and Improve their Attitudes towards STEM? *International Journal of Science Education*, *40*(11), 1263–1283. doi: https://doi.org/10.1080/09500693.2018.1473659.

Wilson, M. (2014). Quality Educational Research Outputs and Significance of Impact: Enduring Dilemma or Stimulus to Learning Transformations between Multiple Communities of Practice? *Hillary Place Papers*, *1*, 1–15. http://hpp.education.leeds.ac.uk/wp-content/uploads/sites/131/2014/03/HPP2013-1Michael_W.pdf.

Winarno, N., Rusdiana, D., Susilowati, E., & Afifah, R. M. A. (2020). Implementation of Integrated Science Curriculum: A Critical Review of the Literature. *Journal for the Education of Gifted Young Scientists, 8*(2), 795–817.

Xu, M., Williams, P. J., Gu, J., & Zhang, H. (2019). Hotspots and Trends of Technology Education in the International Journal of Technology and Design Education: 2000–2018. *International Journal of Technology and Design Education, 30*(2). doi: https://doi.org/10.1007/s10798-019-09508-6.

Zhou, B. (2020). View of Effectiveness of a Pre-College Stem Outreach Program. *Journal of Higher Education Outreach and Engagement, 24*(3), 61–72. https://openjournals.libs.uga.edu/jheoe/article/view/2452/2591.

Zhou, S. N., Zeng, H., Xu, S. R., Chen, L. C., & Xiao, H. (2019). Exploring Changes in Primary Students' Attitudes towards Science, Technology, Engineering and Mathematics (Stem) across Genders and Grade Levels. *Journal of Baltic Science Education, 18*(3), 466.

Appendix

Abdullah, N., Halim, L., & Zakaria, E. (2014). Vstops: A Thinking Strategy and Visual Representation Approach in Mathematical Word Problem Solving toward Enhancing Stem Literacy. *Eurasia Journal of Mathematics, Science and Technology Education, 10*(3), 165–174. doi: https://doi.org/10.12973/eurasia.2014.1073a.

Adnan, M., Abdullah, J. M., Ibharim, L. F. M., Hoe, T. W., Janan, D., Abdullah, N., Idris, N., Wahab, A. S. A., Othman, A. N., Hashim, M. E. A., Said, N. M., Adnan, R., Yahaya, S., Amin, N., Noh, M. A. M., Sufa'at, N. I., Abdullah, R., Yusof, Y., Reduaan, Z. A. M., … Baharudin, N. F. A. (2019). Expanding Opportunities for Science, Technology, Engineering and Mathematics Subjects Teaching and Learning: Connecting through Comics. *Malaysian Journal of Medical Sciences, 26*(4), 127–133. doi: https://doi.org/10.21315/mjms2019.26.4.15.

Adnan, M., Ayob, A., Ong, E. T., Ibrahim, M. N., Ishak, N., & Sheriff, J. (2016). Memperkasa Pembangunan Modal Insan Malaysia Di Peringkat Kanak- Kanak: Kajian Kebolehlaksanaan Dan Kebolehintegrasian Pendidikan Stem Dalam Kurikulum Permata Negara. *Geografia: Malaysian Journal of Society & Space, 12*(1), 29–36.

Al-Rahmi, W. M., Othman, M. S., & Yusuf, L. M. (2015). The Role of Social Media for Collaborative Learning to Improve Academic Performance of Students and Researchers in Malaysian Higher Education. *International Review of Research in Open and Distance Learning, 16*(4), 177–204. doi: https://doi.org/10.19173/irrodl.v16i4.2326.

Alias, N., DeWitt, D., & Siraj, S. (2014). An Evaluation of Gas Law Webquest Based on Active Learning Style in a Secondary School in Malaysia. *Eurasia Journal of Mathematics, Science and Technology Education, 10*(3), 175–184. doi: https://doi.org/10.12973/eurasia.2014.1074a.

Ariffin, S. A., Side, S. F., & Mutalib, M. F. H. (2018). A Preliminary Investigation of Malaysian Student's Daily Use of Mobile Devices as Potential Tools for Stem in a Local University Context. *International Journal of Interactive Mobile Technologies, 12*(2), 80–91. doi:10.3991/ijim.v12i2.8015.

Azman, M. N. A., Sharif, A. M., Parmin, Balakrishnan, B., Yaacob, M. I. H., Baharom, S., Zain, H. H. M., Muthalib, F. H. A., & Samar, N. (2018). Retooling Science Teaching on Stability Topic for Stem Education: Malaysian Case Study. *Journal of Engineering Science and Technology, 13*(10), 3116–3128.

Baharin, N., & Kamarudin, N. (2018). STEM Asean Project to Promote 21st Century Teaching and Learning. *Learning Science and Mathematics Journal, 13*, 98–114.

Bahri, N. M., Suryawati, E., & Osman, K. (2014). Students' Biotechnology Literacy: The Pillars of STEM Education in Malaysia. *Eurasia Journal of Mathematics, Science and Technology Education*, *10*(3): 195–207. doi: https://doi.org/10.12973/eurasia.2014.1074a.

Bahrum, S., & Ibrahim, M. N. (2018). Kebolehgunaan Modul 'Steam' Dalam Pengajaran Dan Pembelajaran Pendidikan Seni Visual Sekolah Rendah (Usability of 'Steam' Module in Teaching and Learning of Visual Arts Education in Primary School). *KUPAS SENI: Jurnal Seni Dan Pendidikan Seni (Journal of Art and Art Education)*, *6*(1), 65–79.

Balakrishnan, B., & Azman, M. N. A. (2017). Professionals Back to School – An Engineering Outreach Programme: A Case Study in Malaysia. *Journal of Engineering Science and Technology*, *12*(10), 2640–2650.

Chew, C. M., & Idris, N. (2016). Form Four Science Students' Perceptions of the Quality of Learning Experiences Provided by Assessments in Stem Related Subjects. *International Journal of Assessment and Evaluation in Education*, 5, 50–56.

Fadzil, H. M., & Saat, R. M. (2014). Enhancing STEM Education during School Transition: Bridging the Gap in Science Manipulative Skills. *Eurasia Journal of Mathematics, Science and Technology Education*, *10*(3), 209–18. doi: https://doi.org/10.12973/eurasia.2014.1071a.

Gelamdin, R. B., & Daniel, E. G. S. (2016). Malaysian Secondary School Students' Knowledge and Interest in Biotechnology: A Case Study. *Journal of Science and Mathematics Education in Southeast Asia*, *39*(1), 24–42.

Gopal, K., Salim, N. R., & Ayub A. F. M. (2019). Perceptions of Learning Mathematics among Lower Secondary Students in Malaysia: Study on Students' Engagement Using Fuzzy Conjoint Analysis. *Malaysian Journal of Mathematical Sciences*, *13*(2), 165–185.

Halim, L., Rahman, N. A., Wahab, N., & Mohtar, L. E. (2018). Factors Influencing Interest in Stem Careers: An Exploratory Factor Analysis. *Asia-Pacific Forum on Science Learning and Teaching*, *19*(2).

Hashim, H., Ali, M. N., & Shamsudin, M. A. (2018a). Enhancing an Entrepreneurial Mindset in Secondary School Students by Introducing the Green-Stem Project via the Integration of the 6E Instructional Model. *Journal of Science and Mathematics Education in Southeast Asia*, *41*(1), 173–192.

Hashim, H., Ali, M. N., & Samsudin, M. A. (2018b). Nurturing Habits of Mind (HOM) through Thinking Based Learning (TBL) in Doing Science Technology, Engineering and Mathematics (STEM) Project. *EDUCATUM Journal of Science, Mathematics and Technology*, *5*(2), 7–18. doi: https://doi.org/10.37134/ejsmt.vol5.2.2.2018.

Hasran, U. A., Jalil, N. F. A., Din, R., Daud, W. R. W., & Noor, S. F. M. (2019). Pendidikan Teknologi Multidisiplin: Mengenali Sel Fuel Dengan Pendekatan Pembelajaran Berasaskan Permainan. *Asean Journal of Teaching and Learning in Higher Education (AJTLHE)*, *11*(2), 36–54. http://ejournals.ukm.my/ajtlhe/article/view/37904.

Hassan, M. N., Abdullah, A. H., Ismail, N., Suhud, S. N. A., & Hamzah, M. H. (2018). Mathematics Curriculum Framework for Early Childhood Education Based on Science, Technology, Engineering and Mathematics (STEM). *International Electronic Journal of Mathematics Education*, *14*(1), 15–31. doi: https://doi.org/10.12973/iejme/3960.

Hoon, T. S., Singh, P., Han, C. T., Nasir, N. M., Rasid, N. S. B. M., & Zainal, N. B. (2020). An Analysis of Knowledge in STEM: Solving Algebraic Problems. *Asian Journal of University Education*, *16*(2), 131–140. doi: 10.24191/AJUE.V16I2.10304.

Huri, N. H. D., & Karpudewan, M. (2019). Evaluating the Effectiveness of Integrated Stem-Lab Activities in Improving Secondary School Students' Understanding of Electrolysis. *Chemistry Education Research and Practice*, *20*(3), 495–508. doi: https://doi.org/10.1039/c9rp00021f.

Hussin, H., Jiea, P. Y., Rosly, R. N. R., & Omar, S. R. (2019). Integrated 21st Century Science, Technology, Engineering, Mathematics (Stem) Education through Robotics Project-Based Learning. *Humanities and Social Sciences Reviews*, 7(2), 204–211. doi:10.18510/hssr.2019.7222.

Hussin, H., Kamal, N., Ibrahim, M. F. (2019). Inculcating Problem Solving and Analytical Skills in Stem Education Practices: The Crystal Initiatives. *International Journal of Innovation, Creativity and Change*, 9(6), 260–272.

Jajuri, T., Hashim, S., Ali, M. N., & Abdullah, S. M. S. (2019). The Implementation of Science, Technology, Engineering and Mathematics (Stem) Activities and Its Effect on Student's Academic Resilience. *Asia Pacific Journal of Educators and Education*, 34, 153–166. doi: 10.21315/apjee2019.34.8.

Jamel, F. M., Ali, M. N., & Ahmad, N. J. (2019). The Needs Analysis in Game-Based Stem Module Development for KSSM Science Teachers. *International Journal of Recent Technology and Engineering*, 8(3), 6622–6628. doi: 10.35940/ijrte.C5655.098319.

Junaidi, M. I. M., Zain, S. M., Hamid, R., Basri, N. E. A., Suja, F., Osman, S. A., & Yasin, R. M. (2017). Initial Impact Analysis of Stem Educational Model by Innovative Sustainable Home (Ig-Home) Using the Rasch Measurement Model. *Journal of Engineering Science and Technology*, 12(Special Is), 91–98.

Kamal, N., Ibrahim, M. F., & Huddin, A. B. (2019). Evaluation of Scratch Programming Mentoring Program amongst Primary School Students. *International Journal of Innovation, Creativity and Change*, 9(6), 243–259.

Kamal, A. A., Shaipullah, N. M., Truna, L., Sabri, M., & Junaini, S. N. (2020). Transitioning to Online Learning during Covid-19 Pandemic: Case Study of a Pre-University Centre in Malaysia. *International Journal of Advanced Computer Science and Applications*, 11(6), 217–123. doi: https://doi.org/10.14569/IJACSA.2020.0110628.

Khalid, M., & Embong, Z. (2019). Sources and Possible Causes of Errors and Misconceptions in Operations of Integers. *International Electronic Journal of Mathematics Education*, 15(2). doi: https://doi.org/10.29333/iejme/6265.

Khalid, S. N., Musa, M., Rahmat, F., Mohamed, N. A., & Mat, N. A. A. (2019). Pembangunan Dan Penilaian Modul Pengajaran Stem Dalam Bidang Statistik Dan Kebarangkalian Dalam Kssm Matematik Tingkatan Dua (Development and Evaluation of Stem Teaching Module in the Field of Statistics and Probability in Form Two Mathematics KSSM). *Journal of Quality Measurement and Analysis*, 15(2), 25–34.

Kok, K. H., Yasin, R. M., & Amin, L. (2018). Pelaksanaan Pendekatan Interdisiplin Dengan Bioteknologi Dalam Sains Tambahan (Interdisciplinary Approach with Biotechnology in Additional Science). *Jurnal Pendidikan Malaysia*, 43(2), 49–59. doi: https://doi.org/10.17576/jpen-2018-43.02-05.

Ku, P. L., & Lim, S. C. J. (2018). Perlaksanaan Dan Keberkesanan Kaedah Lattice Dalam Pengajaran Kemahiran Matematik: Satu Kajian Kes Di Sekolah Rendah. *Online Journal for TVET Practitioners*, 3(Special Issue), 219–231.

Kumar, J. A., & Al-Samarraie, H. (2018). MOOCS in the Malaysian Higher Education Institutions: The Instructors' Perspectives. *Reference Librarian*, 59(3), 163–177. doi: 10.1080/02763877.2018.1458688.

Kumar, J. A., & Al-Samarraie, H. (2019). An Investigation of Novice Pre-University Students' Views towards MOOCS: The Case of Malaysia. *Reference Librarian*, 60(2): 134–147. doi: 10.1080/02763877.2019.1572572.

Lapawi, N., & Husnin, H. (2020). The Effect of Computational Thinking Module on Achievement in Science. *Science Education International*, 31(2). doi: https://doi.org/10.33828/sei.v31.i2.

Lee, S. Y. (2019). Promoting Active Learning and Independent Learning among Primary School Students Using Flipped Classroom. *International Journal of Education, Psychology and Counseling*, 4(30), 324–341. http://www.ijepc.com/PDF/IJEPC-2019-30-05-23.pdf.

Loh, S. L., Pang, V., & Lajium, D. (2019). Evaluation of the Implementation of a Science Project through the Application of Integrated Stem Education as an Approach. *International Journal of Education, Psychology and Counseling*, 4(33), 263–283. doi: https://doi.org/10.35631/IJEPC.4330021.

Loh, S. L., Pang, V., & Lajium, D. (2020). A Case Study of Teachers' Pedagogical Content Knowledge in the Implementation of Integrated Stem Education. *Jurnal Pendidikan Sains Dan Matematik Malaysia*, 10(1), 49–64. doi: https://doi.org/10.37134/jpsmm.vol10.1.6.2020.

Majid, N. A. A., & Majid. N. A. (2018). Augmented Reality to Promote Guided Discovery Learning for STEM Learning. *International Journal on Advanced Science, Engineering and Information Technology*, 8(4–2), 1494–1500. doi: 10.18517/ijaseit.8.4-2.6801.

Matawali, A., Bakri, S. N. S., Jumat, N. R., Ismail, I. H., Arshad, S. E., & Din, W. A. (2019). The Preliminary Study on Inverted Problem-Based Learning in Biology among Science Foundation Students. *International Journal of Evaluation and Research in Education*, 8(4), 713–718. doi: 10.11591/ijere.v8i4.20294

Meerah, T. S. M., Harail, M. F. A., & Halim, L. (2012). Malaysian Secondary School Students' Knowledge and Attitudes towards Biotechnology. *Journal of Baltic Science Education*, 11(2), 153–163.

Meng, C. C., Idris, N., & Eu, L. K. (2014). Secondary Students' Perceptions of Assessments in Science, Technology, Engineering, and Mathematics (STEM). *Eurasia Journal of Mathematics, Science and Technology Education*, 10(3), 219–127. doi: https://doi.org/10.12973/eurasia.2014.1070a.

Najib, S. A. M., Mahat, H., & Baharudin, N. H. (2020). The Level of Stem Knowledge, Skills, and Values among the Students of Bachelor's Degree of Education in Geography. *International Journal of Evaluation and Research in Education*, 9(1), 69–76. doi: 10.11591/ijere.v9i1.20416.

Nasri, N. M., Nasri, N., & Talib, M. A. A. (2020). Towards Developing Malaysia STEM Teacher Standard: Early Framework. *Universal Journal of Educational Research*, 8(7), 3077–3084. doi: 10.13189/ujer.2020.080736.

Nawi, N. D., Phang, F. A., Mohd-Yusof, K., Rahman, N. F. A., Zakaria, Z. Y., Hassan, S. A. H. B. S., & Musa, A.N. (2019). Instilling Low Carbon Awareness through Technology-Enhanced Cooperative Problem Based Learning. *International Journal of Emerging Technologies in Learning*, 14(24), 152–166. doi: https://doi.org/10.3991/ijet.v14i24.12135.

Ng, W. (2019). Affective Profiles of Year 9/10 Australian and South East Asian Students in Science and Science Education. *EURASIA Journal of Mathematics, Science and Technology Education*, 16(1). doi: https://doi.org/1029333/ejmste/110782.

Noh, A. M., & Khairani, A. Z. (2020). Validating the S-Stem among Malaysian Pre-University Students. *Jurnal Pendidikan IPA Indonesia*, 9(3), 421–429. doi:10.15294/jpii.v9i3.24109.

Ompok, C. S. C. C., Ling, M. T., Abdullah, S. N. M. @ S., Tambagas, M., Tony, E. E., & Said, N. (2020). Mentor-Mentee Programme for STEM Education at Preschool Level. *Southeast Asia Early Childhood Journal*, 9(1), 1–14.

Ong, E. T., Ayob, A., Ibrahim, M. N., Adnan, M., Shariff, J., & Ishak, N. (2016). The Effectiveness of an In-Service Training of Early Childhood Teachers on Stem Integration through Project-Based Inquiry Learning (PIL). *Journal of Turkish Science Education,* *13*(Special issue), 44–58. doi: 10.12973/tused.10170a.

Ong, K. J., Chou, Y. C., & Yang, D. Y. (2019). The Impact of Science Fair on the Students' Engagement, Capacity, Continuity, and Motivation towards Science Learning. *Jurnal Pendidikan Sains Dan Matematik Malaysia, 9*(1), 1–12. doi: https://doi. org/10.37134/jpsmm.vol9.1.1.2019.

Ong, E. T., Safiee, N., Jusoh, Z. M., Salleh, S. M., & Noor, A. M. H. M. (2017). STEM Education through Project-Based Inquiry Learning: An Exploratory Study on Its Impact among Year 1 Primary Students. *Jurnal Pendidikan Sains Dan Matematik Malaysia, 7*(2), 43–51. doi: https://doi.org/10.37134/jpsmm.vol7.2.4.2017.

Ong, S. L., & Wah, J. L. P. (2019). UCTS Foundation Students' Perception towards Arduino as a Teaching and Learning Tool in STEM Education. *E-Bangi, 16*(3), 1–21.

Pang, Y. J., Hussin, H., & Ahmad, S. S. S. (2018). Integrated Robotics STEM Curriculum towards Industry 4.0. *International Journal of Human and Technology Interaction (IJHaTI), 2*(2), 17–24. https://journal.utem.edu.my/index.php/ijhati/article/ view/3700/3479.

Pang, Y. J., Hussin, H., Tay, C. C., & Ahmad, S. S. S. (2019). Robotics Competition-Based Learning for 21st Century STEM Education. *Journal of Human Capital Development (JHCD), 12*(1), 83–100. https://journal.utem.edu.my/index.php/jhcd/ article/view/5389/3854.

Pang, Y. J., Tay, C. C., Ahmad, S. S. S., & Ng, K. T. (2019). Promoting Students' Interest in STEM Education through Robotics Competition-Based Learning: Case Exemplars and the Way Forward. *Learning Science and Mathematics Journal, 14,* 107–121. http:// www.recsam.edu.my/sub_LSMJournal/images/docs/2019/2019_8_PYJ_107121_ Final.pdf.

Radzi, N. A. M., & Sulaiman, S. (2018). Measuring Students' Interest towards Engineering in Technical School: A Case Study. *Journal of Technology and Science Education, 8*(4), 231–237. doi: 10.3926/jotse.369.

Rasid, N. S. M., Nasir, N. A. M., Singh, P., & Han, C. T. (2020). STEM Integration: Factors Affecting Effective Instructional Practices in Teaching Mathematics. *Asian Journal of University Education, 16*(1), 56–69. doi: doi:10.24191/ajue.v16i1.8984.

Rasul, M. S., Zahriman, N., Halim, L., & Rauf, R. A. A. (2018). Impact of Integrated STEM Smart Communities Program on Students Scientific Creativity. *Journal of Engineering Science and Technology, 13*(Special Issue on ICITE 2018), 80–89.

Razali, S. N. R., Hamid, M. E. A., Hamid, N. S. A., Puteh, M., Kamil, W. M. A. W. M., Zainuddin, Z., Awang, R., Azhar, M. A. D. M., Mastor, M. Z. S., Sazalli, N. A. H., Zainon, O., Sjarif, N. N. A., Omar, M. S. S., Law, E. L. C., & Bannister, N. (2020). Astronomy Outreach Programs with STEM Ambassadors under the C3aol Project. *Jurnal Kejuruteraan, 3*(1), 43–49. doi: https://doi.org/10.17576/jkukm-2020-si3(1)-07.

Razali, F., Manaf, U. K. A., & Ayub, A. F. M. (2020). STEM Education in Malaysia towards Developing a Human Capital through Motivating Science Subject. *International Journal of Learning, Teaching and Educational Research, 19*(5), 411–422. doi: 10.26803/IJLTER.19.5.25.

Razali, F., Manaf, U. K. A., Talib, O., & Hassan, S. A. (2020). Motivation to Learn Science as a Mediator between Attitude towards STEM and the Development of STEM Career Aspiration among Secondary School Students. *Universal Journal of Educational Research, 8*(1A), 138–146. doi: 10.13189/ujer.2020.081318.

Romero, M., Arnab, S., Smet, C. D., Mohamad, F., Minoi, J. L, & Morini, L. (2019). Assessment of Co-Creativity in the Process of Game Design. *Electronic Journal of e-Learning, 17*(3), 199–206. doi: 10.34190/JEL.17.3.003.

Rosli, R., Abdullah, M., Siregar, N. C., Hamid, N. S. A., Abdullah, S., Gan, K. B., Halim, L., Daud, N. M., Bahari, S. A., Majid, R. A., & Bais, B. (2020). Student Awareness of Space Science: Rasch Model Analysis for Validity and Reliability. *World Journal of Education*, *10*(3), 170. doi: https://doi.org/10.5430/wje.v10n3p170.

Saleh, S., Muhammad, A., & Abdullah, S. M. S. (2020). STEM Project-Based Approach in Enhancing Conceptual Understanding and Inventive Thinking Skills among Secondary School Students. *Journal of Nusantara Studies (JONUS)*, *5*(1), 234–254. doi: https://doi.org/10.24200/jonus.vol5iss1pp234-254.

Saleh, S., & Rahman, M. A. A. (2016). A Study of Students' Achievement in Algebra: Considering the Effect of Gender and Types of Schools. *European Journal of STEM Education*, *1*(1), 19–26. doi: https://doi.org/10.20897/lectito.201603.

Samat, N. A. A., Ibrahim, N. H., Surif, J., Ali, M., Abdullah, A. H., Talib, C. A., & Bunyamin, M. A. H. (2019). Chem-A Module Based on STEM Approach in Chemical Bond. *International Journal of Recent Technology and Engineering*, *7*(6), 700–710.

Samsudin, M. A., Jamali, S. M., Zain, A. N. M., & Ebrahim, N. A. (2020). The Effect of STEM Project Based Learning on Self-Efficacy among High-School Physics Students. *Journal of Turkish Science Education*, *17*(1), 94–108. doi: 10.36681/tused.2020.15.

Samsudin, M. A., Zain, A. N. M., Jamali, S. M., & Ebrahim, N. A. (2018). Physics Achievement in STEM Project Based Learning (PBL): A Gender Study. *Asia Pacific Journal of Educators and Education*, *32*, 21–28. doi: https://doi.org/10.21315/apjee2017.32.2.

Shahali, E. H. M., L. Halim, L., Rasul, M. S., Osman, K., & Zulkifeli, M. A. (2017). STEM Learning through Engineering Design: Impact on Middle Secondary Students' Interest towards STEM. *Eurasia Journal of Mathematics, Science and Technology Education*, *13*(5), 1189–1211. doi: 10.12973/eurasia.2017.00667a.

Shahali, E. H. M., Halim, L., Rasul, M. S., Osman, K., & Arsad, N. M. (2019). Students' Interest towards Stem: A Longitudinal Study. *Research in Science and Technological Education*, *37*(1), 71–89. doi: 10.1080/02635143.2018.1489789.

Shahali, E. H. M., Halim, L., Rasul, M. S., Osman, K., Ikhsan, Z., & Rahim, F. (2015). Bitara-STEM™ Training of Trainers' Programme: Impact on Trainers' Knowledge, Beliefs, Attitudes and Efficacy towards Integrated STEM Teaching. *Journal of Baltic Science Education*, *14*(1), 85–95.

Sharif, A. M., Azman, M. N. A., Balakrishnan, B., Yaacob, M. I. H., & Zain, H. H. M. (2018). The Development and Teachers' Perception on Electromagnet Teaching Aid: Magnobolt. *Journal Pendidikan IPA Indonesia*, *7*(3), 252–258. doi: 10.15294/jpii.v7i3.13491.

Shukri, A. A. M., Che Ahmad, C. N., & Daud, N. (2019). Implementing a Celik STEM Module in Empowering Eighth-Graders' Creative Thinking. *International Journal of Education, Psychology and Counseling*, *4*(32), 219–37. doi: https://doi.org/10.35631/ijepc.4320021.

Siew, N. M. (2017). Fostering Students' Scientific Imagination in STEM through an Engineering Design Process. *Problems of Education in the 21st Century*, *75*(4), 375–93.

Siew, N. M., & Ambo, N. (2018). Development and Evaluation of an Integrated Project-Based and STEM Teaching and Learning Module on Enhancing Scientific Creativity among Fifth Graders. *Journal of Baltic Science Education*, *17*(6), 1017–1033. doi: https://doi.org/10.33225/jbse/18.17.1017.

Siew, N. M., & Ambo, N. (2020). The Scientific Creativity of Fifth Graders in a STEM Project-Based Cooperative Learning Approach. *Problems of Education in the 21st Century*, *78*(4), 627–643. doi: https://doi.org/10.33225/pec/20.78.627.

Siew, N. M., Goh, H., & Sulaiman, F. (2016). Integrating Stem in an Engineering Design Process: The Learning Experience of Rural Secondary School Students in an Outreach Challenge Program. *Journal of Baltic Science Education, 15*(4), 477–493.

Singh, P., Teoh, S. H., Cheong, T. H., Md Rasid, N. S., Kor, L. K., & Md Nasir, N. A. (2018). The Use of Problem-Solving Heuristics Approach in Enhancing STEM Students Development of Mathematical Thinking. *International Electronic Journal of Mathematics Education, 13*(3). doi: https://doi.org/10.12973/iejme/3921.

Tan, W. L., Samsudin, M. A., Ismail, M. E., & Ahmad, N. J. (2020). Gender Differences in Students' Achievements in Learning Concepts of Electricity via Steam Integrated Approach Utilizing Scratch. *Problems of Education in the 21st Century, 78*(3), 423–448. doi: https://doi.org/10.33225/pec/20.78.423.

Tan, G. G. S., Shakawi, A. M. H. A., & Azizan, F. L. (2017). Relationship between Students' Diagnostic Assessment and Achievement in a Pre-University Mathematics Course. *Journal of Education and Learning, 6*(4), 364–371. doi: https://doi.org/10.5539/jel.v6n4p364.

Tauro, F., Cha, Y., Rahim, F., Sattar Rasul, M., Osman, K., Halim, L., Dennisur, D., Esner, B., & Porfiri, M. (2017). Integrating Mechatronics in Project-Based Learning of Malaysian High School Students and Teachers. *International Journal of Mechanical Engineering Education, 45*(4), 297–320. doi: 10.1177/0306419017708636.

Wan Husin, W. N. F., Mohamad Arsad, N., Othman, O., Halim, L., Rasul, M. S., Osman, K., & Iksan, Z. (2016). Fostering Students' 21st Century Skills through Project Oriented Problem Based Learning (POPBL) in Integrated STEM Education Program. *Asia-Pacific Forum on Science Learning and Teaching, 17*(1), Article 3. https://www.eduhk.hk/apfslt/download/v17_issue1_files/fadzilah.pdf

Yasin, R. M., Amin, L., & Kok, K. H. (2018). Interdisciplinary M-Biotech-Stem (MBS) Module for Teaching Biotechnology in Malaysia. *K-12 STEM Education, 4*(2), 341–362. doi: https://doi.org/10.14456/k12stemed.2018.6.

Yeoh, M. (2018). Problem-Based Learning (PBL) among Malaysian Teachers: An Evaluation on the In-Service Training of Facilitation Skills. *Learning Science and Mathematics Online Journal (SEAMEO RECSAM), 13*(December), 59–72.

Yung, O. C., Junaini, S. N., Kamal, A. A., & Md Ibharim, L. F. (2020). 1 Slash 100%: Gamification Of Mathematics with Hybrid Qr-Based Card Game. *Indonesian Journal of Electrical Engineering and Computer Science, 20*(3), 1453–1459. doi:10.11591/ijeecs.v20.i3.pp1453-1459.

Zaharin, N. L., Sharif, S., Singh, S. S. B., Talin, R., Mariappan, M., Mohanaraj, N., Jusup, Y., & Suppiah, P. (2019). Promoting Students' Interest, Attitude and Intrinsic Motivation towards Learning Stem through Minimalist Robot Education Programme. *International Journal of Service Management and Sustainability, 4*(1), 41–66. doi: https://doi.org/10.24191/ijsms.v4i1.8054.

Zainal, N. F. A., Din, R., Majid, N. A. A., Nasrudin, M. F., & Rahman, A. H. A. (2018). Primary and Secondary School Students Perspective on KOLB-Based STEM Module and Robotic Prototype. *International Journal on Advanced Science, Engineering and Information Technology, 8*(4–2), 1394–1401. doi: 10.18517/ijaseit.8.4-2.6794.

Zainal, N. F. A., Din, R., Nasrudin, M. F., Abdullah, S., Rahman, A. H. A., Abdullah, S. N. H. S., K. Ariffin, A. Z., Jaafar, S. M., & Majid, N. A. A. (2018). Robotic Prototype and Module Specification for Increasing the Interest of Malaysian Students in STEM Education. *International Journal of Engineering and Technology (UAE), 7*(3), 286–290. doi: doi:10.14419/ijet.v7i3.25.1758.

11 STEM/STEAM education research in South Korea

Oksu Hong

Introduction: Science, Mathematics, Informatics, and STEAM education in Korea

With the fourth industrial revolution bringing about fundamental changes in the industrial structure and reshaping the future of jobs, current education systems aim to cultivate future talent with complex problem solving, critical thinking, and creativity (World Economic Forum [WEF], 2016). In particular, Science, Technology, Engineering, and Mathematics (STEM) education has gained enormous global traction in recent years, with new jobs expected to be created in the STEM field due to the rapid development of science and technology such as artificial intelligence (AI), big data, and Internet of Things (IoT).

The Korean government enacted the Science, Mathematics, and Informatics Education Promotion Act (hereinafter the "Act") in 2018 intending to establish a support system for the education of Science, Mathematics, and Informatics (SMI), which are core subjects associated with the Fourth Industrial Revolution (4IR). The Act aims at nurturing future talents with convergent skills to lead the era of 4IR by promoting activities such as research on the evaluation of the contents and methods of SMI education, development of textbooks and educational materials, and the operation of teacher professional development programs. In line with the Act, the Korean Ministry of Education (MOE) and the Korea Foundation for the Advancement of Science and Creativity (KOFAC) developed a "Master plan for Science, Mathematics, Informatics, and Convergence Education" to be applied for five years, which started in 2020. This plan, which pursues the vision of developing future global leaders with competencies and qualities required by the intelligent information society, contains mid-to-long-term policy directions and concrete tasks. Driven by this master plan, educational institutions will step up their efforts to promote Korean STEM education.

One of the most salient Korean STEM education features is its emphasis on Science, Technology, Engineering, Arts, and Mathematics (STEAM) education to highlight STEM's integration-oriented approach. STEAM education aims to foster students' ability to solve real-world problems. Most real-world problems cannot be solved with the knowledge of a single subject but can be solved by connecting and utilising knowledge gathered from various academic

DOI: 10.4324/9781003099888-11

fields. Thus, STEAM education aims at the natural integration needed in solving real-world problems and requires the integration of two or more subjects among S, T, E, A, and M (Korea Foundation for the Advancement of Science and Creativity [KOFAC], 2016). In this context, convergence is being used as a keyword to highlight the characteristics of Korean STEAM education with an integration-oriented approach. The Korean government has continually driven STEAM education since 2011; Korean STEAM education is positively evaluated because it has contributed to enhancing students' self-directed learning abilities, science preferences, interests in science and mathematics, and self-efficacy, naturally achieving a paradigm shift from traditional lecture-centred to student-centred classes (Ministry of Education [MOE], 2018).

Korea is now in the process of a national curriculum revision scheduled for 2022. At present, subjects are organised under the 2015 national curriculum, which was established to foster creative-integrated talents with humanistic imagination, scientific creativity, and upright characters who can create new knowledge and values by integrating various knowledge (Kim et al., 2015). Two main directions of change set by the 2022 revision of the national curriculum are (1) The implementation of the high school credit system aimed at allowing students more academic freedom and supporting aptitude-based personalised education, including the reorganisation of science, mathematics, and informatics curricula; and (2) Curriculum operation taking account of the spread of non-face-to-face education due to COVID-19. The demand for personalised, customised, problem-based, and collaborative forms of learning, which have been set as the direction of educational transformation for the future generations, has been accelerated by COVID-19, and there are heated discussions about the curricular contents and learning processes to realise this demand.

In the sections that follow, trends of the STEM/STEAM education in Korea will be examined, revolving around the policy and research promoted by the KOFAC as the government agency responsible for the STEM/STEAM education in Korea.

Research in science education

The 2015 national curriculum currently in place is a competency-based curriculum, in which overarching and subject-specific competencies are identified. The national science curriculum was developed to cultivate five competencies: Scientific thinking ability, scientific inquiry ability, scientific problem-solving ability, scientific communication ability, scientific participation, and lifelong learning ability (MOE, 2015; see Table 11.1 for the science subjects). The 2015 science curriculum differentiates itself from the previous ones in the following aspects: (1) Prioritization of affective domain as a learning objective; (2) Consideration of lifelong learning of science for competencies; and (3) Addition of the subject "Integrated Science" centring on core concepts of natural phenomena (Song, 2020).

Table 11.1 Subjects of the science curriculum of the 2015 national curriculum in Korea

		Elective-centred curriculum		
School-level	*Common curriculum*	*Common courses*	*General electives course*	*Career-related electives course*
Elementary School	Science			
Middle School	Science			
High School		• Integrated Science • Science Inquiry Lab	• Physics I • Chemistry I • Life Science I • Earth Science I	• Physics II • Chemistry II • Life Science II • Earth Science II • Science History • Life and Science • Convergence Science

With the aim of setting guidelines for future-oriented science education to develop creative-integrated talents, KOFAC released the Korean Science Education Standards for the Next Generation (KSES-NG) after five years of preparation (2014–2019) and developed the model Tree of Scientific Literacy (ToSL) (Song et al., 2019). The ToSL model has three roots representing scientific literacy: Competence, knowledge, and participation and action. Each dimension consists of domains and subdomains (Figure 11.1). The ToSL model defines scientific literacy, the goal of science education, as "the attitudes and abilities as democratic citizens to participate in and act for solving personal as well as social problems using science-related competencies and knowledge".

Currently, MOE-led research for curriculum revision is underway under the heading of "research on future-oriented science curriculum organisation to cope with post-COVID challenges" to redesign science subjects taking into account the scientific competencies and deriving teaching and learning strategies and assessment methods tailored to the post-COVID world (Jeong et al., 2021). This research project kicks off of the full-fledged science curriculum development research towards the publication of the new national curriculum in 2022, to be applied from 2025 onwards.

In Korea, all science education policies are established based on the Five-Year Master Plan for Science Education. The current master plan (2020–2024) proposes six action plans and 16 priority tasks to attain the three objectives of science education compacting the foundation, enjoying the cutting edge, and leading the future (MOE, 2020). One of the core objectives of this master plan is expanding the "intelligent science lab" to all primary and secondary schools by 2024 as an online and offline space in which scientific inquiry can be made using cutting-edge scientific technologies, such as AI, Augmented Reality (AR)/Virtual Reality (VR), and IoT. Given that the abilities to predict the future based on data and solve problems in the real world are the core

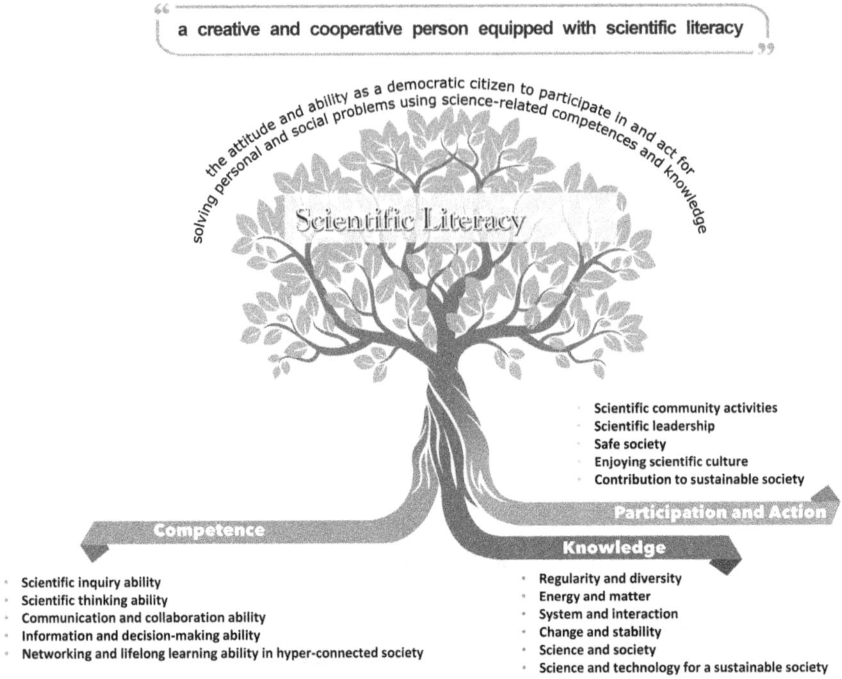

Figure 11.1 ToSL model of "scientific literacy" in KSES-NG (Song et al., 2019).

competitiveness in the intelligent information society, the intelligent science lab supports scientific inquiry activities using real data through an online scientific inquiry platform named "Intelligent Science Lab-ON". The online platform has the following features: (1) Scientific inquiry data platform; (2) IoT sensor and device information sharing system for data measurement and collection; (3) Data-based inquiry activity sharing network; (4) Teaching and learning platform and scientific inquiry software platform interface (Lim et al., 2020).

In order to empower science teachers in their research and practice, KOFAC has been supporting the project "IDEA (Interest, Development, Engagement, and Association) Science Teacher Research Groups" since 2016. The IDEA Science Teacher Research Groups contribute to science teachers' competence self-empowerment through research-based development of teaching materials on various topics and using them in classrooms and the spread of exemplary classroom activities across the country by reinforcing teachers' networks. In 2020, 46 IDEA Science Teacher Research Groups were supported, and a range of science class models have been developed, such as advanced technology utilisation (e.g., AI, big data, and IoT), empirical problem-solving, real-life problem-solving (e.g., SSI: Socio-Scientific Issue), and student-centred collaboration (KOFAC, 2021).

To further disseminate innovation models of science education, many projects were conducted at the school level. First, the "creative-integrated-science-lab schools" operated in schools across the country to innovate school science labs to where students can conduct active and inquiry-based learning, encompassing not only scientific inquiry activities (e.g., experiments, investigations, and debates), but also project activities (e.g., problem-based learning and creative problem-solving). The creative-integrated science lab has sections of physical spaces for presentation, learning, experimentation, discussion, preparation, exhibition, and creation to reinforce students' collaboration and scientific inquiry. From 2021 onwards, the creative-integrated science lab will be operated in tandem with the intelligent science lab. According to the survey results analysing 120 schools that operated as creative-integrated science lab schools from 2017 to 2019, the creative-integrated science labs had a positive effect in increasing interest in scientific inquiry and improving science inquiry ability, presentation ability, and collaboration skills (Son et al., 2020).

The "student-participatory-science-class leading school" is also being operated across the country to encourage active student participation in science classes. There are six class types proposed for this model: Discussion/debate, investigation/presentation, production/expression, project, experiment/observation, and problem-solving. The process of student-initiated learning is also included in performance assessment so that various facets of each student's changes and growth are documented, and appropriate feedback is provided (Kim, Yong-jin et al., 2019). In 2020, the COVID-19 pandemic forced schools to switch to online classes for the student participation model. Science teachers, who participated in the "Forum on student participation online science classes in the post-COVID era" organised by KOFAC suggested that student-participatory science classes might as well be continued through the revitalisation of intramural and extramural learning communities, expansion of digital literacy education for teachers and students, sharing-centred teacher training, and proactive use of EdTech (KOFAC, 2020a).

The "science core schools" operating since 2009 are high schools specialising in science and mathematics. The curriculum operation of science core schools is characterised by the emphasis on curricula that form the basis for furthering creativity by strengthening students' proficiency in mathematics, physics, chemistry, life science, and earth science. Currently, there are 124 science core schools across the country in 2021. According to the results of a survey conducted in 2020, students enrolled in science core school were satisfied with programs such as science and mathematics experiential activities, club activities, and research & education (R&E), and graduates acknowledged that the curricula and educational activities in science core schools helped them learn their majors at the post-secondary level (Choi et al., 2021).

For cutting-edge technologies to be reflected in primary and secondary science education, the space of science education needs to go beyond the school context to society. To this end, KOFAC-led "STAR (school teacher and research) Bridge Centre" model development research is underway intending to transform science education by connecting schools to universities, businesses, and local communities and establishing a collaboration network between teachers and science and technology professionals. Specifically, STAR Bridge Centres aim to develop future-oriented teaching materials, run teacher training programs, and develop an AI-based learning content management system, under joint collaborative research with the institute of science and technology, science museums and centres, and EdTech companies (Im et al., 2020).

As examined above, a variety of national-level research and practice activities are being carried out in science education, and the Korean Science Education Index (KSEI) has recently been developed to comprehensively examine and assess the quality of science education (Hong et al., 2021c). Based on a Delphi survey conducted with 25 experts in science education, science education policy, and science and technology, the categories of the KSEI were grouped into input, process, and outcome. The "input" category consists of the factors "student characteristics", "teacher characteristics", and "education infrastructure"; the "process" category comprises "science curriculum operation", "science contents development and program operation", and "teacher professional development program operation"; and the "outcomes" category includes "cognitive achievements", "affective achievements", "scientific competencies", and "participation and action." The KSEI will be used for the annual collection of data for a comprehensive assessment of the current state of science education, and the results of data analysis will be used for setting up a national science education policy.

Research in Mathematics education

The 2015 Korean Mathematics National Curriculum was developed to help students understand and acquire knowledge and skills of mathematics, as well as cultivate six mathematical competencies: Problem-solving, reasoning, creativity and integration, communication, information processing, and attitude and practice (MOE, 2015). The revised curriculum differentiates itself from the previous ones in the following aspects: (1) Conceptualization of mathematical competencies, (2) Reduction of students' learning burden, (3) Emphasis on the affective aspects, (4) Utilisation of technology, and (5) Reorganisation of real-life-centred statistics (Lee, 2020).

Among the mathematics curriculum subjects presented in Table 11.2, Basic Mathematics and Mathematics in Artificial Intelligence (AI) for high school were added in 2020 as newly developed subjects. First, Basic Mathematics contains basic-level content for students who have difficulty learning high school mathematics (Kwon et al., 2020). It was designed to substitute high school mathematics subjects in consideration of the school type and the student's level of achievement and can be understood as an implementation of a curriculum tailored to students. Mathematics in Artificial Intelligence is a future-oriented new subject in which students can learn and experience how mathematical concepts are applied to AI. This subject consists of four domains: AI and mathematics, data presentation, classification and prediction, and optimisation (Lee et al., 2020).

Table 11.2 Subjects of the mathematics curriculum of the 2015 national curriculum in Korea

School-level	Common curriculum	Elective-centred curriculum		
		Common courses	General electives course	Career-related electives course
Elementary School	Mathematics			
Middle School	Mathematics			
High School		• Mathematics	• Mathematics I • Mathematics II • Calculus • Probability and Statistics	• Geometry • Practical Mathematics • Mathematics for Economics • Mathematics Project • Basic Mathematics • Mathematics in Artificial Intelligence (AI)

Currently, MOE-led research for curriculum revision is underway under the heading of "Research on future-oriented mathematics curriculum organisation to cope with post-COVID challenges" to derive the vision and content of mathematics education in consideration of the changes in society such as changes in the demographic structure, the development of science and technology, and the advent of a hyper-connected society (Lee et al., 2021). This research will serve as the basis for the national mathematics curriculum to be released in 2022.

In Korea, all mathematics education policies are established based on the Five-Year Master Plan for Mathematics Education. The current master plan (2020–2024) proposes five action plans and 14 priority tasks to attain the three objectives of "mathematics classroom without students giving up on math", "fostering real-life problem-solving skills", and "developing core math talents" (MOE, 2020). This plan includes strategies for the "AI Mathematics Learning System" capable of AI-based diagnosis of individual achievement levels and learning deficiency elements and providing learning content to make up for the deficiencies. It aims to contribute to the realisation of education for all students' mathematical literacy and the improvement of equity in education by providing the necessary AI system free of charge to all schools. To this end, a KOFAC-led research project is underway to develop a model that supports customised education tailored to each student's achievement level by collecting and analysing big data reflecting each student's cognitive and affective characteristics (Park et al., 2021).

MOE and KOFAC have been supporting primary and secondary mathematics education using technology. First, AlgeoMath (www.algeomath.kr) as –

the name suggests – covering all mathematics from algebra to geometry, is a software program that allows the user to draw graphs and charts according to the input equations. Features such as geometric, algebraic, block coding, data utilisation, and 3D functions are available on the user interface of the online platform, and teaching and learning materials and teacher training programs are continuously being developed (Hong et al., 2021a). The Toctoc Math Expedition (www.toctocmath.kr) ["toctoc" is a homonym meaning smart or knock-knock], an AI-based elementary mathematics class support system using a gamification approach, was developed to minimise math learning deficiencies in lower graders. The baseline number sense test and basic arithmetic test are used for diagnosing students' level of mathematics learning, and customised learning contents recommended by AI based on the scores are provided to students. It has been used nationwide since 2020 and has been particularly efficient in improving students' learning in the underachiever group, as per the developments' intentions (Hong et al., 2021b).

To support engagement in math for all, out-of-school facilities are necessary. In Korea, base-type math cultural centres (differently named in each city, e.g., Math Park in Busan and Math Is Our Mate in Daejeon), are being operated in Seoul and Gyeongnam. In the basic research for the establishment of math cultural centres, the following roles they need to play were proposed: (1) Inform the essence of mathematics; (2) Show cutting-edge mathematics; (3) Exhibit and experience mathematics; (4) Communicate mathematics; (5) Motivate students to learn mathematics; (6) Recognise the importance and necessity of mathematics; (7) Experience the joy of mathematics; and (8) Provide information about mathematics (Do et al., 2015). Currently, math cultural centres are increasingly playing an important role in math communication and education.

Research in Informatics education

One of the most salient features of the 2015 national curriculum is the emphasis on software education, which reflects the need to foster computational thinking: A core competence indispensable to future generations. This demand was reflected in the 2015 national curriculum by making informatics a compulsory subject at the secondary level (see Table 11.3 for main changes in the 2015 informatics curriculum). Computational thinking refers to mental skills for understanding and solving problems in everyday life and various academic fields based on computer science's basic concepts and principles (MOE, 2015). Although researchers and educators have agreed on the significance of computational thinking, many teachers faced the challenge of the lack of assessment standards and tasks. KOFAC developed an assessment model of computational thinking using the evidence-centred design (ECD) approach and derived four key assessment areas (i.e., decomposition, abstraction, algorithm, and automation) and three supporting assessment areas (i.e., analysis, communication, and collaboration) (Park et al., 2017).

Table 11.3 Changes in the 2015 informatics curriculum

School-level (Application year)	Main changes		Contents
	Before	*After*	
Elementary school (2019–Present)	ICT unit in practical arts (12 hours)	Software education in practical arts (over 17 hours)	• Problem-solving process, algorithm, programming concepts
Middle school (2018–Present)	Informatics * elective subject	Informatics (over 34 hours) * compulsory subject	• Information ethics • Problem-solving based on CT • Developing algorithms and programming (Intermediate level focusing on the block-based coding)
High school (2018–Present)	Informatics * advanced elective subject	Informatics * general elective subject	• Designing algorithms and programs with various fields (Advanced level focusing on the text-based coding)

KOFAC and the Korea Information Science Education Foundation (KiSEF) jointly developed a next-generation software education standard model by itemising the contents at the elementary, middle school, and high school levels considering the sequential flow, continuity, and connectivity (Kim, KapSu et al., 2019). As shown in Figure 11.2, information and computer literacy were

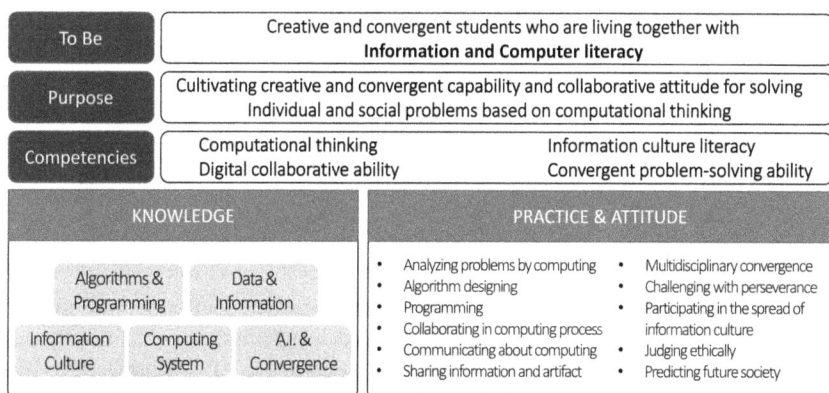

To Be	Creative and convergent students who are living together with **Information and Computer literacy**	
Purpose	Cultivating creative and convergent capability and collaborative attitude for solving Individual and social problems based on computational thinking	
Competencies	Computational thinking Digital collaborative ability	Information culture literacy Convergent problem-solving ability

KNOWLEDGE	PRACTICE & ATTITUDE
Algorithms & Programming / Data & Information / Information Culture / Computing System / A.I. & Convergence	• Analyzing problems by computing • Algorithm designing • Programming • Collaborating in computing process • Communicating about computing • Sharing information and artifact • Multidisciplinary convergence • Challenging with perseverance • Participating in the spread of information culture • Judging ethically • Predicting future society

Figure 11.2 Framework of the next generation informatics education standard model (Kim, KapSu et al., 2019).

presented as big ideas, and computational thinking, information culture literacy, digital collaborative ability, and convergent problem-solving ability were proposed as core competencies. The knowledge dimension was classified into five domains of information culture, data and information, algorithms and programming, a computing system, and AI and convergence, whereby "AI and convergence" were newly added to the 2015 national curriculum.

With AI gaining traction as a new core issue, KOFAC conducted a basic study on primary and secondary level AI education. In the study, it defined AI literacy, AI-driven problem-solving ability, and AI-driven ethical thinking as core competencies that students should be equipped with, and proposed "Understanding of AI", "AI and data", "AI algorithms", "Application of AI", and "AI and social impact" as five learning areas (Yoo et al., 2020). An extension of these core competencies and learning areas, Artificial Intelligence (AI) Basics was added in 2020 as a newly developed subject for a national curriculum for high school education (Kim, Jaehyun et al., 2020; see Table 11.4 for the details).

In 2020, the MOE set up the First Master Plan for Informatics Education and proposed three objectives (i.e., systemisation of curriculum, the establishment of future-oriented educational infrastructure, and development of AI talents), seven action plans and 17 priority tasks (MOE, 2020). This plan includes an AI education platform development project for all applicable to pre-collegiate students, university students, and the general public who want to learn the principles of AI, machine learning, and data science. The AI education platform will gradually enter the full-fledged implementation stage starting from 2021, beginning with 2,000 AI/Software leading schools across the country.

With increasing emphasis on informatics education at primary and secondary education levels, there has been a growing interest in improving teachers' competencies. KOFAC developed a teacher training curriculum optimised for the school levels (elementary, middle, and high schools) and achievement levels (basic, general, and advanced) and is operating training sessions in both offline and online settings (KOFAC, 2020b). Furthermore, KOFAC is currently proceeding with the SWEET (SW Education for all Elementary Teachers) project to bring elementary teacher education colleges into offering software-related compulsory courses to provide pre-service teachers with adequate professional competencies. A significant positive correlation was observed in the questionnaire survey between the respondents' self-efficacy level and the number of software-related courses they completed (Kim Ja Mee et al., 2020).

Research in STEAM education

STEAM education in Korea aims to enhance student's interest and understanding in science and technology (S&T) and foster integrative thinking and problem-solving. It pursues three core competencies: (1) "Knowledge and information processing": The ability to process and utilise knowledge and information in various areas to reasonably resolve problems; (2) "Creative thinking": The ability to create new things by using the knowledge, skills, and experience

Table 11.4 Contents of artificial intelligence basics for high school (Kim, Jaehyun et al., 2020)

Subject area	Concept	Generalized knowledge	Content elements
Understanding AI	AI and society	AI is a core technology that constitutes the Fourth Industrial Revolution and is leading social and occupational changes.	• Concept and characteristics of AI • The evolution of AI technology and social changes
	AI and agents	AI solves problems by perceiving, learning, reasoning about, and acting on its environment in the form of an intelligent agent.	• The concept and role of intelligent agents
Principles and applications of AI	Perception	Intelligent agents interact with the world through visual, auditory, and other means of perception.	• Sensors and perception • Computer vision • Voice recognition and language comprehension
	Search and reasoning	Search explores diverse paths to problem-solving knowledge of the world is represented in a structured way and used to find solutions by reasoning.	• Problem-solving and search • Knowledge representation and reasoning
	Learning	Learning in AI is the automatic creation of a model from data, which handles classification, clustering, prediction, and other tasks.	• Machine learning concepts and applications • Deep learning concepts and applications
Data and machine learning	Data	Data is used to implement a machine learning model. It can be classified into structured and unstructured data.	• Data attributes • Structured vs. unstructured data
	Machine learning model	A machine learning model is implemented by defining intelligent problems, preparing data necessary for problem-solving, training the model, and testing.	• Classification model • Machine learning model implementation
Social impact of AI	The impact of AI	AI can have positive and negative impacts on individual lives and society at large.	• Solving social problems • Data bias
	AI ethics	AI ethics is a set of values and a mode of behaviour required of society members to ensure proper utilisation of AI.	• Ethical dilemmas • Social responsibility and fairness

in various areas of expertise in an integrated manner based on basic knowledge; and (3) "Communication": The ability to effectively express one's thoughts and feelings in various situations and attentively listen to others' opinions and respect them (MOE, 2018). STEAM education has been implemented since 2011 as part of government-led policy projects to develop S&T talents. MOE has supported primary and secondary schools by implementing the following four action plans to promote STEAM education: (1) Integration of STEAM education into educational settings by specifying the STEAM-related contents in the objectives of the curriculum; (2) Promotion of STEAM education by supporting the STEAM leading schools and teacher research groups; (3) Development of STEAM programs by supporting collaborative work among universities, enterprises, research institutions, and other professionals and wide distribution of the programs to schools; and (4) Exploration of improvement and expansion strategies for STEAM education by supporting policy research on STEAM education.

In the meantime, MOE released the Master Plan for Science, Mathematics, and Informatics Education along with the Five-Year Convergence Master Plan (2020–2024). The Convergence Master Plan reflects the intention to strengthen interdisciplinary convergence education, centring on science, mathematics, and informatics, to develop creative-integrated talents with core competencies required for coping with the future society based on the achievements and efforts of STEAM education thus far. In setting up this master plan, a KOFAC-led study was conducted, and the direction and core competencies of convergence education derived in the study are presented in Figure 11.3 (Park et al., 2019).

The 2015 national curriculum emphasised fostering creative-integrated talents equipped with core competencies, such as the 4C (i.e., Critical thinking,

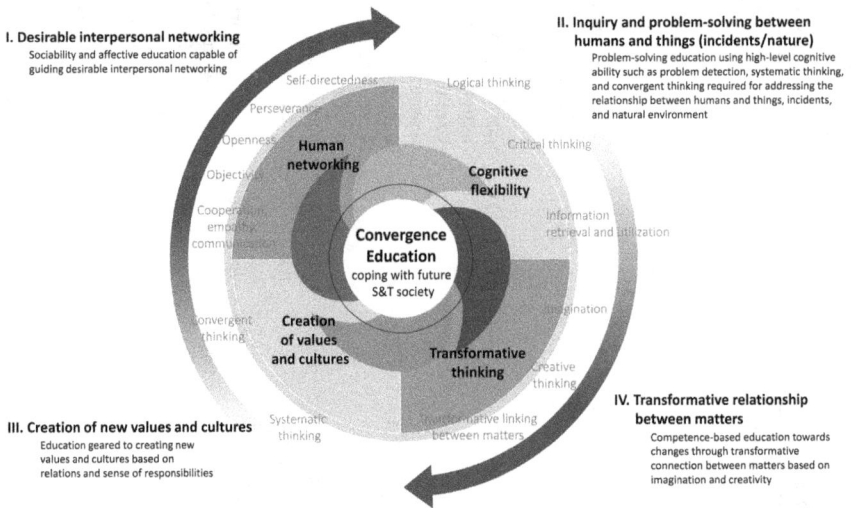

Figure 11.3 Direction and core competencies of convergence education 2030 (Park et al., 2019).

Communication, Collaboration, and Creativity), cultivated through STEAM education that pursues interdisciplinary convergence and connection between academia and everyday life. STEAM education was reflected in the national science curriculum, and STEAM-type activities were included in the science textbooks (MOE, 2015). For STEAM education to be effectively implemented in school, adequate assessment of learning outcomes is indispensable. In this respect, the "STEAM Learning Assessment Model" was proposed considering the objectives of STEAM education, which the authors defined as enhancing students' interest and understanding in the S&T fields, fostering S&T-based convergence thinking, and real-world problem-solving ability (Ryu et al., 2019). The proposed model designed as a competence-centred modular assessment model links STEAM classroom learning and assessment and is capable of both process-centred and product-centred assessments (Figure 11.4).

For the direction of the convergence education 2030 (Figure 11.3) to be put in place, a spatial model is necessary. Thus, a study on future-oriented space for convergence education was conducted, which proposed four modular spaces: (1) Personalisation space for information processing; (2) Understanding space for self-directed activities and presentation; (3) Collaboration space for group activities and project implementation; and (4) Sharing space for sharing experiences and connecting to the other spaces (Hong et al., 2019). This study emphasised the design of convergence education space tailored to participants – specifically, play-driven convergence education for elementary school students, process-oriented convergence education for middle school students, convergence

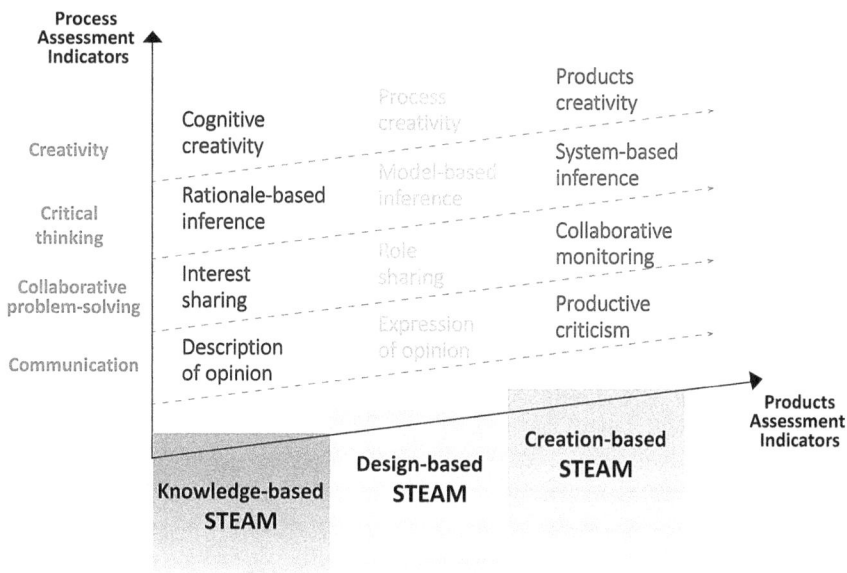

Figure 11.4 STEAM learning assessment model (Ryu et al., 2019).

education based on the basic knowledge of various fields for high school students, convergence education leading to the creation of new values based on the majors and area expertise for university students and postgraduate/doctoral students, and convergence education based on social issues and personal interests for the general citizens.

Given that a variety of STEAM education materials have been developed under the initiatives of MOE and KOFAC, a study performed a comprehensive analysis of these materials in conjunction with the subject areas of the 2015 national curriculum and compiled a content map (Kim et al., 2017). According to the study findings, the main focus of STEAM education materials in Korea is on extracurricular activities such as creative experience for elementary school students, free semester system for middle school students, and after-school activities for high school students. In particular, students 'satisfaction with the topic-dependent convergence, high-tech product application, science-art convergence, and future job-linked programs developed by experts in various fields was high. In recent years, to actively introduce STEAM education in regular classes, programs linked to national curriculum are also being developed. All the materials developed to date are posted on the STEAM education website (https://steam.kofac.re.kr/).

Conclusions: Future directions for STEM/STEAM education

In the preceding sections, the research trends of STEM/STEAM education in Korea have been examined, focusing on the results of related policy projects and studies carried out by KOFAC. The STEM/STEAM education in Korea can be characterised as follows: (1) Systematisation of core competencies, objectives, content arrangement, and teaching methods for each subject of SMI as specified in the national curriculum; (2) Establishment of mid-to-long-term vision and future-oriented priority tasks according to national-level master plans; and the (3) Enhancement of the quality of education by improving teacher empowerment and developing high-quality contents. This section gives a brief overview of the future directions of STEM/STEAM education in Korea based on the research results yielded so far. The directions proposed here are mainly based on the vision of convergence education described in the master plan for convergence education (2020–2024) (MOE, 2020), and the objectives pursued by the SMI education in the AI age proposed by Hong (2020).

Firstly, it is necessary to strengthen basic education in science, mathematics, and informatics and restore student interest and confidence. "Convergence" is the most important keyword in our times caught in tremendous civilisational changes across the industrial fields and social (living) domains due to the rapid development of AI technology. However, "domain knowledge" is essential in bringing about convergence-based AI-driven transformative changes in society's various domains. Such domain knowledge stems from a solid basic education in science, mathematics, and informatics at the primary and secondary levels.

Furthermore, despite Korean students' high scientific and mathematical literacy achievements, they tend to show low interest and confidence (KICE, 2020). According to the results of a survey on teachers' perceptions, "science-related activities out of school", "student-centred research projects", "ICT-related materials", and "performance assessment" were found to have positive effects on students' positive experiences. Therefore, these approaches should be actively brought to bear in all school classes.

Secondly, it is necessary to establish a more flexible educational environment with an integrated approach in dealing with offline school environments and online platform environments. As explored above, in Korea, a range of studies are being conducted on the intelligent science lab, AI-driven math learning support systems, AI education platforms, and future-oriented convergence education spaces and they are sequentially applied to the educational settings. To implement future-oriented education, learning environments must be redesigned, and cutting-edge technologies and tools must be employed to transform education. Changes in the educational environment will be accelerated to meet the growing demand for EdTech-driven personalised, problem-based, and collaborative learning due to the ongoing non-face-to-face education situation caused by COVID-19. In line with this trend, it is necessary to establish a flexible on/offline learning environment that enables students and teachers to undertake various educational activities, coupled with the upgrading of teaching materials and enhancement of teacher empowerment to keep abreast with the changing environments.

Thirdly, an ecosystem needs to be established for science, mathematics, and informatics education through the concerted efforts of individuals, schools, and society. Korea's school-centred, well-organised education system has been considered desirable for Korean primary and secondary education. However, considering the rapid changes in science, technology, and society, a more flexible curriculum operation is required, and a sustainable education ecosystem needs to be established to ensure efficient and organic utilisation of various educational resources through cooperation with universities, companies, public institutions, science museums, math cultural centres, and other related facilities. In light of this movement, it is under review to recognise the credit hours gained in informal settings in the high school credit system, which will be progressively applied from 2022 onwards. Like the aforementioned STAR (School Teacher and Research) Bridge Centre, various extramural institutions can develop forward-looking educational materials, operate teacher training programs, and improve primary and secondary education quality.

When future generations have jobs that do not currently exist, talented leaders who are to implement innovative transformations in many fields of society through the convergence across fields will be needed. Currently, not a convergence occurring in an individual's brain, but a convergence occurring in close collaboration with experts from various disciplines is changing the world by creating new values and cultures. This chapter looks forward to the Korean STEM/STEAM education contributing to the training of talents who will contribute to people and society's well-being.

References

Choi, Jae-Hyeok et al., *A report on consulting and research on 2020 science core schools* (Seoul: KOFAC, 2021).

Do, Young Hae et al., *Basic research of establishing the center for Mathematical culture* (Seoul: KOFAC, 2015).

Hong, Jun Euy et al., *A study on the plan to create future convergence education space* (Seoul: KOFAC, 2019).

Hong, Oksu, "The latest trends and orientations in science, mathematics, and informatics education in the AI era." *The Science & Technology* 609 (2020): 25–29.

Hong, Oksu et al., *AlgeoMath 4th year development research report* (Seoul: KOFAC, 2021a).

Hong, Oksu et al., *A study on the development and operation of AI-based elementary mathematics class support system* (Sejong: Ministry of Education, 2021b).

Hong, Oksu et al., *Development of a comprehensive index for Korean science education* (Seoul: KOFAC, 2021c).

Im, Sung Min et al., *Research on the establishment of the center for supporting science education* (Seoul: KOFAC, 2020).

Jeong, Dae Hong et al., *Research on future-oriented science curriculum organization to cope with post-COVID challenges* (Sejong: Ministry of Education, 2021).

Korea Foundation for the Advancement of Science and Creativity (KOFAC). *Online forum on student participation online science classes in the post-COVID era* (Seoul: KOFAC, 2020a).

Korea Foundation for the Advancement of Science and Creativity (KOFAC). *2020 KOFAC science education data book* (Seoul: KOFAC, 2021).

Korea Foundation for the Advancement of Science and Creativity (KOFAC). *2020 software education annual report* (Seoul: KOFAC, 2020b).

Korea Foundation for the Advancement of Science and Creativity (KOFAC). *Introduction to STEAM education* (Seoul: KOFAC, 2016).

Korea Institute for Curriculum and Evaluation (KICE). *The Trends in International Mathematics and Science Study (TIMSS): finding from TIMSS 2019 for Korea* (Jincheon: KICE, 2020).

Kim, Jaehyun et al., *A study on the development of <Artificial Intelligence Basics> subject according to the 2015 national curriculum* (Sejong: Ministry of Education, 2020).

Kim, KapSu et al., *A study on the development of the standard of software education for the next generation* (Seoul: KOFAC, 2019).

Kim, Kyung Ja et al., *2015 revised national curriculum general outline development* (Ministry of Education, 2015).

Kim, Ja Mee et al., *A study on effectiveness of support program for strengthening software education in teacher training college* (Seoul: KOFAC, 2020).

Kim, Yong-jin et al., *Development of STEAM contents map and analyses of STEAM programs in 2017* (Seoul: KOFAC, 2017).

Kim, Yong-jin et al., *Guidelines for student-participatory-science-class leading school* (Seoul: KOFAC, 2019).

Kwon, Oh Nam et al., *A study on the development of <Basic Mathematics> subject according to the 2015 national curriculum* (Sejong: Ministry of Education, 2020).

Lee, Kyeong Hwa et al., *Research on future-oriented mathematics curriculum organization to cope with post-COVID challenges* (Sejong: Ministry of Education, 2021).

Lee, Sang-gu et al., *A study on the development of <mathematics in artificial intelligence (AI)> subject according to the 2015 national curriculum* (Sejong: Ministry of Education, 2020).

Lee, YoungJun, "Tasks and orientation of national science curriculum," in *The reflections on the Korean national curriculum reform*, ed. Young Dal, Cho (Gyeonggi: Kyoyook Book, 2020), 5–27.

Lim, Wan Chul et al., *A study on the development of software for scientific inquiry* (Seoul: KOFAC, 2020).

Ministry of Education (MOE). *2015 revised national curriculum* (Sejong: Ministry of Education, 2015). http://ncic.re.kr/mobile.dwn.ogf.inventoryList.do#.

Ministry of Education (MOE). *The master plan for STEAM education (2018–2022)* (Sejong: Ministry of Education, 2018).

Ministry of Education (MOE). *The master plan for science, mathematics, informatics, and convergence education (2020–2024)* (Sejong: Ministry of Education, 2020).

Park, Hyunju et al., *A study for establishing the master plan of convergence education* (Seoul: KOFAC, 2019).

Park, Man Goo et al., *Planning and research on the AI mathematics learning system to support math learning* (Seoul: KOFAC, 2021).

Park, Seman et al., *Designing an evidence-centered assessment model of computational thinking in elementary and secondary education* (Seoul: KOFAC, 2017).

Ryu, Suna et al., *A study on the assessment method for STEAM education* (Seoul: KOFAC, 2019).

Son, Jeongwoo et al., *A study on the development of creative-integrated-science-lab schools through analysis of satisfaction* (Seoul: KOFAC, 2020).

Song, Jinwoong et al., *Scientific literacy for all Koreans: Korean science education standards for the next generation* (Seoul: KOFAC, 2019).

Song, Jinwoong, "Tasks and orientation of national science curriculum," in *The reflections on the Korean national curriculum reform*, ed. Young Dal, Cho (Gyeonggi: Kyoyook Book, 2020), 157–180.

World Economic Forum (WEF). *The future of jobs: Employment, skills and workforce strategy for the fourth industrial revolution* (Geneva: World Economic Forum, 2016), http://www3.weforum.org/docs/WEF_Future_of_Jobs.pdf.

Yoo, JeongSu et al., *An exploratory study on the content system of artificial intelligence education in elementary and secondary schools* (Seoul: KOFAC, 2020).

12 Understanding STEM integration in Singapore using complex, persistent, and extended problems

Aik-Ling Tan and Roxanne Lau Shu Xin

Introduction

The complexities of real-world problems in the 21st century demand knowledge and skills across different disciplines to solve. Consequently, this has resulted in a push for integration in learning, particularly in STEM fields (Science, Technology, Engineering, and Mathematics), since the key challenges faced by the world are both caused by and can be solved by advances in science and technology. It is believed that understanding the connections in epistemic, conceptual, and social aspects across disciplines will enable learners to generate more creative, inclusive, and sustainable solutions to grand challenges such as climate change, poverty, clean water, and transport (United Nations Sustainable Development Goals, 2015). As such, moving integrated STEM education forward would require the collaboration of all stakeholders (Reeves, 2021) such as science teachers, mathematics teachers, technology teachers, policymakers, engineers, and industries. The question remains on the form, characteristics, and responsibilities of multi-party collaboration of STEM education stakeholders.

For teachers working directly with students in schools to learn through integrated STEM experiences, it is crucial for them to be able to work in teams, identify meaningful problems, design engaging activities, facilitate inquiry, comment on students' designs, and assess students' learning among other duties they need to perform. Teachers, therefore, play multiple roles of being a learner, risk-taker, inquirer, curriculum designer, negotiator, collaborator, and teacher (Slavit et al., 2016). As such, the support that teachers need includes access to material and time resources, clear instructional frameworks to design STEM activities, teacher professional development, and engagement in professional learning communities. This chapter aims to share insights from a research conducted in a Singaporean classroom where students worked on an integrated STEM problem to illustrate possible student-teacher interactions, particularly with regard to questioning, during integrated STEM learning.

As learners respond and interact with teachers, their ideas, and chosen strategies, students' understanding of the intentions of the learning activities, their abilities to engage in problem identification, problem solving, and gathering relevant evidence to persuade determine the success of the learning experience.

DOI: 10.4324/9781003099888-12

Assumptions that students understand problem solving and are able to make the right connections across disciplines by simply working on STEM problem solving could possibly be unwarranted. In fact, Honey et al. (2014) recommended that teachers need to help students make the needful connections explicitly. The tenets of effective STEM pedagogical practices include a focus on inquiry within contextualised settings, drawing relevance to students' everyday lives, creating opportunities for students to be engaged in argumentation, reasoning, and digital learning (McDonald, 2016). The process of inquiry is appropriate to science, mathematics, engineering and technology as it involves raising questions or identifying problems, investigations, evaluating, crafting explanations, optimising conditions, and using evidence to inform decisions (National Research Council [NRC], 2012). Questions and answers during whole-class discussions are one way to help students make connections within and between disciplines. Questioning is a complementary process that supports discussion and argumentation of ideas (Chin & Osborne, 2008), both of which are needful processes in problem solving. As such, this research examines the questions raised when students are engaged in STEM problem solving.

A common character of STEM education used by STEM education researchers is the amalgamation of real-world contexts, challenging problem solving, teamwork, and constructing multidisciplinary connections, which constitutes deeper learning as explained in Otto et al. (2020). Further, these pedagogical methods match the goals of the Education 4.0 initiative introduced in the recent report by the World Economic Forum, where it states that "interactive methods that promote the critical and individual thinking (are) needed in today's innovation-driven economy" (World Economic Forum, 2020). Given the popularity of integrated STEM practices, it is time to better understand students' STEM learning experiences. In this chapter, we focus on the following research questions as we examine a lesson where students engage in an integrated STEM task centred on a problem. The research questions are:

1 How are different types of questions utilised by teachers as part of their instructional practices during problem-centric STEM lessons?
2 What questions do students raise when they are engaged in integrated STEM problem solving?

Literature review

The S-T-E-M quartet

Identification of suitable problems is the first step to learning about STEM problem solving. Locating the problems in appropriate curricular frameworks, choosing the right strategies and understanding how students interact with the problems and the subsequent learning from solving the problem are necessary second steps since they are likely to influence students' learning experiences and hence motivation in learning. Dierdorp et al. (2014) analysed students' responses to a questionnaire after planned lesson units in mathematics, statistics, and science and found that students need to experience coherence between different disciplines to enable them to make connections. They proposed two forms of curriculum coherence, namely connections between different disciplines and connections between what is learnt and its cohesion to the real world. Meaningful scientific concepts, problems, and contexts are characterised by usefulness, motivation, application, authenticity, and connections. Star et al. (2014) worked with students on learning mathematics through a technology-based activity. Using a pre- and post-test design, they found that motivational experiences need to be aligned to the developmental level of students. Further, Barrett et al. (2014) proposed a structure for interdisciplinary curriculum in which there is usually a portion of 20 to 25 minutes where the teacher engages students in whole-class question-and-answer followed by a 30 to 35 minutes of hands-on activity. Time is also allocated for testing the structures built. Through this curriculum structure of question-answers to familiarise themselves with relevant concepts, application through hands-on activities and testing of their prototypes, they found improvements in test scores for all questions. English and King (2015) worked with grade four students on engineering design and found that application of disciplinary knowledge occurred most frequently in the last two phases of engineering design, namely design evaluation and redesign. This implies that activities need to lead students to these stages before the science and engineering ideas can emerge. Finally, Crismond and Adam (2012) proposed performance indicators for evaluating design and argued for the need for problem framing over problem solving, doing research rather than skipping research, idea fluency rather than idea scarcity, and the need to balance benefits and trade-offs rather than ignoring them. The literature reviewed above identified the features of curriculum frameworks that support integrated STEM. The various research findings point to the need for positioning integrated STEM problems within a thoughtful curriculum framework that has the following features: (1) Space for students to make connections within and beyond disciplines, (2) Appropriate scaffolds for learners of different levels of readiness, (3) Opportunities for questions and answer, (4) Avenue for testing and refining design and prototype, and (5) Immersion in a fluid and flexible environment that encourages idea fluency and trade-offs.

Considering the features of integrated STEM curriculum frameworks reviewed earlier, the STEM Quartet instructional framework proposed by Tan et al. (2019) was chosen as a guide to design activities in this study. This is because the STEM Quartet is centred around a complex, persistent, and extended problem, and involves vertical learning within a discipline as well as horizontal connections between different disciplines. Persistent problems are problems that recur often and hence they often serve as "organising points for knowledge" (Bereiter, 1992, p. 346). A persistent problem is a challenge that is non-routine and are problems of explanations that are applicable across different contexts. Complex problems refer to challenges or issues that demand knowledge and skills from across different disciplines to generate solutions. Knowledge and skills from a single discipline are insufficient to design solutions that are holistic and convincing. Finally, extended problems refer to problems that require learners to work on the issues over a prolonged period of time. Learners can engage with the problems from three to about ten hours. The intention to develop an engineer's strand of thought amongst students is aligned with the objective of STEM education spelt out in Bryan et al. (2016). Furthermore, the positive correlation between science inquiry and engineering design is explained in Purzer et al. (2015), which mentioned that the two entities both foster "learning by doing". As such, the STEM activity emphasises more on the Science and Engineering aspects.

In planning and crafting integrated STEM problems that are persistent, complex, and extended, it is important to be mindful of the context in which problems and subsequent solutions are located. Winter (1968) proposed a four-part problem-solving pattern of *Situation, Problem, Solution,* and *Evaluation.* A proposed solution to a specific problem is subjected to evaluation and subsequent refinement based on evaluation by specific users. This four-part pattern was adapted by Hoey (2001) when he proposed to replace Solution with Response, arguing that only when the response is positively evaluated will it be considered a solution. Hoey illustrated his idea of "response-positive evaluation becomes solution" using the story of Goldilocks. Goldilocks was lost in the woods (this is the Situation), she became hungry (Problem), she ate Papa Bear's porridge (Response) and found it too hot (negative Evaluation). Then we go back to the problem since Goldilocks is still hungry. When Goldilocks ate Baby Bear's porridge (Response), she found it just right (positive Evaluation). Hence the solution to Goldilocks hunger was to eat Baby Bear's porridge. Hoey's (2001) idea of response-evaluation places the user at the centre of any proposed "solution" or response. A proposed "solution" is only good if the user finds it useful and evaluates it positively. One user may evaluate the response positively while another may evaluate it negatively. As such, evaluation to responses may be different and learners' awareness of specific users' evaluations and concerns can help with improvements of design of proposed ideas.

Questions

Research (e.g., Chin, 2007; Chin & Osborne, 2010; Tan et al., 2017) emphasised the indispensable role of student and teacher questioning in learning science as well as in argumentation. Scientific inquiry is a learning model in which students engage in hands-on activities that allow them to construct knowledge through learning by doing (Satchwell & Loepp, 2002). Questioning plays a crucial part in science discourse and inquiry-based learning (Wells, 2016). Teacher questioning significantly influences students' responses (Lee & Irving, 2018), while student questioning facilitates teaching and learning, and guides students' thinking (Kuhn, 2009).

Studies have shown that effective teachers demonstrate purposeful science discourse skills in teaching inquiry skills and in helping students to construct scientific knowledge (Gillies & Baffour, 2017; Hardy et al., 2010). Dialogic communication and having more student-teacher interactions in classrooms have long been proven to be more effective teaching methods over the conventional mono-logic top-down delivery. Teachers' classroom discourse can be analysed through teacher questioning which determines the structure of classroom talk, scaffolds discussion, and guides students' cognitive process (Chin, 2007). Hardy et al. (2010) also emphasised the strong impact that teacher prompts place on student conceptual understanding and argumentation skills. Therefore, by analysing teachers' questions, we can study how class discussions are initiated (Wells, 2016) as well as how it affects students' argumentation patterns (Webb et al., 2006).

Student questioning also has implications in both teaching and learning of science (Chin & Osborne, 2008). Asking questions helps students to evaluate their understanding (Chin, 2006) and facilitate knowledge construction (Wells, 2016). As part of social negotiation, student questioning during group activities provides insights into students' thought processes and is evidence for the presence of higher-order thinking (Tan et al., 2017). Examining students' arguments show how students make use of evidence and conceptual understanding in questioning assumptions, warranting their ideas, and persuading their ideas to their peers, hence displaying critical thinking (Duschl, 2008). Hence, the role of student questioning in inquiry-based learning is pivotal in that it helps students to steer their learning, and self-review their understanding. For teachers, student questioning can be used as diagnostic tool to assess student understanding or lack thereof. Student questioning can be harnessed to foster productive and purposeful social negotiation patterns to allow for meaningful collaborative and cooperative learning to take place (Chin & Brown, 2002) as well as generate epistemic understanding of an argument (Nam & Chen, 2017).

Research context

In this chapter, we present one case of a larger study in Singapore that targets students' learning experiences with integrated STEM activities focusing on complex, persistent, and extended problems. The 19 grade eight students are from a school in the eastern part of Singapore. The teacher (Ally, pseudonym) in this study is an experienced teacher with more than 15 years of teaching experience and holds a doctoral degree in marine biology. Although the students did not have any formal lessons on photosynthesis, we assumed that they knew plants require sunlight, carbon dioxide and water to photosynthesise, as this concept was taught in primary schools. Prior to teaching the class, Ally had undergone a two-hour professional development session on teaching integrated STEM lessons. During the session, Ally was presented with the resources needed for the STEM activity and she also had the chance to clear her doubts with the researchers. This professional development session was needful to help overcome the issue of the teacher feeling underprepared in STEM classrooms (El-Deghaidy & Mansour, 2015).

The problem that was presented to the students focused on photosynthesis and agricultural engineering as applied to the issue of food self-sustainability in land-scarce Singapore. Singapore has a land area of 728 km^2 and an estimated population density of 8358 per km². The amount of land allocated for agriculture is limited and hence most food in Singapore is imported. This topic was chosen as it is relevant to Singaporean students who are familiar with the fact that there is high land competition in the country. The vulnerability of Singapore as a result of her reliance on imported food became especially apparent with the panic buying situation in many countries as a result of the COVID-19 pandemic. The proposed problem of food self-sustainability is complex, persistent, and extended as it is characterised by having multiple solutions, being able to be applied to different contexts (different countries and different types of food), and is challenging as well as realistic (Tan et al., 2019). With the context of limited land space, students were tasked to design a prototype for a new farming system that occupies as little space as possible, is cost-effective, and is not labour-intensive, without compromising the crop yield.

The Informed Design Learning and Teaching Matrix by Crismond and Adams (2012) present a list of nine informed design patterns: (1) Problem framing; (2) Doing research; (3) Idea fluency; (4) Deep drawing and modelling; (5) Balance benefits and trade-offs; (6) Valid tests and experiments; (7) Diagnostic troubleshooting; (8) Managed and iterative designing; and (9) Reflective design thinking. These behaviours are incorporated with the engineering process as described in English et al. (2016). The STEM activity that was devised incorporated the stages (where possible) of problematising, researching, generating plausible solutions, designing by drawing, constructing models, presenting, reviewing, and evaluating the solutions as suggested by Crismond and Adams (2012) and English et al. (2016).

The activity took three one-hour lessons to complete. Lesson one included the teacher and students describing the context in which the problem is located. Thereafter, students gathered into groups of four to research high-tech farming methods, ask questions (e.g., task-procedural and clarification questions), present their ideas, and decide either on their new farming system or other ways to solve the food sustainability problem collectively as a group. Students also started designing their prototypes by drawing on their activity handouts. In lesson two, students built a model of their prototype using basic materials such as ice-cream sticks, metal wires, straws, wooden sticks, etc. Finally, students presented their prototype drawing and model in lesson three. During peer presentations, students provided critiques to the ideas presented by their peers. Following that, each group reviewed the pros and cons of their prototypes to assess the feasibility of their prototypes.

Data collection

The first lesson was implemented during curriculum time with three researchers as observers. To record the whole class setting, a video camera was set up at the back of the classroom and was directed towards the front of the classroom, capturing Ally, students, and the projected PowerPoint slides. During group discussions, a group of students was randomly chosen, and the students' discussion was video recorded. Data sources for this study included the videos and notes that the students had jotted down during their discussion to consolidate their findings. The videos were observed for Ally's actions including teacher-student interactions. The videos were also transcribed verbatim to analyse the teacher and student talk. The students' notes served as student artefacts that provided insights into the students' research and thought processes.

Data analysis

The goals of the data analysis were to examine (1) The questions used by Ally to achieve the learning intended of an integrated STEM activity, and (2) The students' social negotiation pattern during a group problem-solving activity. In the analysis of the discourse, the various phases of the lesson were first identified and the duration of each phase and the actions of Ally and the students were noted. Finally, key discussions and questions asked in each phase were examined and categorised.

To achieve the first analysis goal, the essential features of an integrated STEM classroom were mapped out based on the framework for essential features of STEM learning by Toh and Tan (2020). Table 12.1 summarised these features. Ally's questions that were both directed at the class and at the group were

Table 12.1 Essential features of an integrated STEM classroom (Toh & Tan, 2020)

Essential feature	Competencies
Problematising	The teacher is able to:
	• Clearly set the context in which the problem is located.
	• Clearly present the proposed problem to the students, showing how the problem is complex, persistent, and extended.
	• Make or facilitate the formation of explicit links that connect the problem to the context.
Group problem solving	• The teacher provides students with an opportunity to work in groups of two or more students to think of plausible solution(s).
	• Group members are assigned clear roles by the teacher or by themselves.
	• Sufficient and suitable facilitation is given by the teacher.
Design process	• Students are given an opportunity to develop their ideas through creating a prototype in various ways such as drawings, building a model, etc.
	• An opportunity is given to students for review and evaluation of the prototype's feasibility, pros, and cons.
Interdisciplinary solutions	• Plausible solution(s) generated by students assimilates two or more STEM disciplines.
	• Clear connections between the solution(s) and the context are made.
	• Epistemic links between the STEM disciplines integrated into the solution(s) are drawn.
	• Clear explanations of how the solution(s) solves the problem in the context are provided.

Table 12.2 Description of the types of teacher's questions analysed in this report (Chin, 2007)

Type of question	Features	Purpose
Pumping	Direct requests made to a prior student's utterance for elaboration on the proposed idea.	To elicit more information from students and encourage them to develop their ideas further.
Constructive challenge	A challenge to a prior utterance to generate reflective thinking, especially when the student's utterance is incorrect or inaccurate.	To encourage students to re-think their ideas or self-correct their responses.
Framing	Overarching questions to structure or outline a new discussion.	To initiate a discussion and stimulate generative thinking amongst students.
Others	Task-relevant questions that do not classify as framing or pumping or constructive challenge questions.	To give instructions and manage behaviour

identified from the transcript. The questions were then categorised based on the different productive questioning approaches as defined by Chin (2007). We focused on examining how Ally adopted pumping, constructive challenge and framing questions (Table 12.2) to demonstrate the essential features of STEM. The three categories of questions were chosen as they were most frequently used questions from our preliminary viewing of the videos. The remaining questions, such as clarification questions were categorised as "others".

To examine students' social negotiation patterns, we analysed students' questioning during the group discussion. We focused on three types of questions that were common in the student's talk (i.e., generic task-procedural questions, specific task-procedural questions, and clarification questions). Questions that do not fall into any of the three categories were classified as miscellaneous. Table 12.3 summarised the various question types. To strengthen the analysis, student's responses and the notes that they have written during their discussion were also studied.

Results and discussion

The 40-minute lesson presented here had three phases: (1) Problematising, (2) Group problem-solving, and (3) Design process. In the problematising phase, there was an eight-minute teacher-facilitated discussion on the problem of food self-sustainability. Ally led the students in co-constructing the implications that land scarcity has on the problem of food production and food sustainability. The discussion between limited land (context) and sustained food production (problem) presented an opportunity for students to make the connections between the context and the problem. Subsequently, the students proceeded to generate

Table 12.3 Description of the types of student's questions analysed

Type of question	Features	Purpose
Generic task-procedural	No evidence or scientific concepts are used in the questioning.	To seek the next step or action needed to carry on with the task.
Specific task-procedural	Evidence, scientific concepts, or context are included in the questioning.	To ask for reasoning or explanation of an aspect needed to proceed with the task.
Clarification	To ask about information that is ambiguous or missing to the questioner.	To seek confirmation or consolidation of information.
Miscellaneous	All other task-relevant questions that do not classify as task-procedural or clarification	Social interaction

plausible solutions in groups of four (group problem-solving). In their groups, students researched high-tech farming methods, asked questions, and engaged in argumentative discussions to present their ideas, ultimately deciding on their new farming system collectively. The process took about 19 minutes. During this group discussion period, the teacher took the role of a facilitator, guiding the groups of students by answering questions when necessary. Upon confirming their new farming system, the students entered the phase of design process by drawing the design of the system on their handout.

Teacher questioning

In the eight-minute problematising phase, the level of interaction between the teacher and students was at the class level before students were divided into their respective groups for group problem-solving and the design process. In the examples that follow, Ally is represented by "T"; a single unidentifiable student speaker is denoted by "S"; multiple unidentifiable student voices are denoted by "SS"; and "xxx" represents inaudible utterance.

Ally's choice of using framing questions to outline a discussion (Excerpt 12.1) rather than directly delivering the content of the topic is strategic in activating generative thinking in students. The framing questions were also presented on PowerPoint slides, acting as objects of inquiry for students to concentrate on and refer to as the discussion followed (Chin, 2007). Among the 38 questions, 35 were asked in the problematising phase and the remaining three questions were directed at a group during group problem-solving. Pumping is the main strategy used by Ally to contextualise the problem, with 68.4% of teacher's questions being pumps (Table 12.4). In Excerpt 12.1, multiple uses of pumping strategies were used by Ally to establish the fact that Singapore has insufficient land for open-field farming. Ally went on to tease out information from students about Singapore's food supply sources and referred to the panic buying situation (Excerpt 12.1). With that, Ally managed to incorporate current real-world situations into the classroom talk, making the problem relatable to the students.

Excerpt 12.1. Where do we get out food supplies?

	Type of teacher's question
T: *However, we are all human beings and we need to eat right?*	Framing
T: *So, where do we get our food supplies?*	Framing
SS: Import.	
T: Okay, we import from other countries. *Can you name me some of the countries we get our food from?*	Pumping
SS: Malaysia. Vietnam. USA. China. India. Thailand.	
T: I think you are more aware now with the recent series of panic buying occurring in Singapore. *Why, why was there panic buying?*	Pumping
T: (gesturing at a student) *Can you share with us?*	Pumping
S: We're scared that there will be a lockdown in Singapore and there will be xxx	
T: We are afraid that there is a lockdown in SG because of the COVID-19 situation and there can be no food coming to Singapore.	

Ally continued to co-construct knowledge about the pertinence of land scarcity on food self-sustainability with the students using Socratic questioning. Socratic questioning is a questioning technique to tease out more information from students by using a series of prompts (Paul & Elder, 2007). Parallel to STEM instructional practices as delineated by Guzey et al. (2016) and Brown et al. (2011), Ally demonstrated the first essential feature of STEM (i.e., problematising) by engaging the class in a discussion that involved making real-world connections. Hence, Ally had performed strategic questioning approaches which helped to "maintain a rigorous, coherent, engaging, and equitable discussion" (Michaels & O'Connor, 2012, p. 4).

In instances where students had made an inaccurate utterance, Ally responded with a constructive challenge instead of directly correcting their answer. Constructive challenge questions took up only 10.5% of all questions, indicating that the students were able to comprehend the lesson well, showed little misconception, or were too shy to voice their ideas. Excerpt 12.2 shows how Ally effectively used constructive challenge and pumping to elicit the idea of vertical farming from the students. Once again, we see Ally reiterating the student's responses by revoicing, this time posing an additional question to challenge the claim. "Revoicing" (Chapin et al., 2009) students' responses to share their ideas

Table 12.4 Analysis of teacher's questions

Type of question	Number of teacher's questions	Proportion of total questions (%)
Pumping	26	68.4
Constructive challenge	4	10.5
Framing	5	13.2
Others	3	7.90
Total	**38**	**100**

Excerpt 12.2. How can we grow food without space?

	Type of teacher's question
T: *How can we solve this problem of food supply in Singapore? Limited food supplies, what can we do?*	Pumping
S: xxx	
T: (gesturing to a student) *What did you say?*	Others
S: We farm by ourselves	
T: We farm by ourselves. We grow crops ourselves. *But no space what, what to do? So how can we grow food without space?*	Constructive challenge
SS: xxx	
T: *Your friend is doing this (gesturing a vertical structure with both hands) what is this?*	Pumping
SS: Vertical farming	

to the class, hence transforming an individual's knowledge into shared common knowledge. Next, Ally posed a constructive challenge to the claim of farming in Singapore (Excerpt 12.2), generating reflective thinking in not just the student who answered, but in the rest of the class as well. This led to more responses from the class, ultimately eliciting "vertical farming" as a solution. Therefore, Ally had demonstrated purposeful teacher questioning which "followed upon a preceding student contribution in a productive way" (Chin, 2007).

Excerpts 12.1 and 12.2 exhibited how purposeful teacher questioning and using real-world linkages are important in facilitating students' understanding of the problem. As the lesson progressed to group problem-solving, Ally assigned roles to each student in the groups – a notetaker, a researcher, and two article-readers. Students were in charge of their own group's discussion where they raised questions and negotiated their ideas with their group mates while Ally moved between groups to scaffold the discussions when necessary, characterising this as cooperative learning (Thibaut et al., 2018). Creating the chance for group discussions presents a stage for students to communicate STEM concepts, construct multidisciplinary connections, and solve problems by working with one another (Stohlmann et al., 2011).

Student questioning

As there were multiple aspects to the overarching problem and ensuing multiple solutions, students were given the autonomy to research on specific areas of focus, discuss, and work in groups to design a new farming system design as a plausible solution to ensure food self-sustainability by maximising crop yield in a limited area of land. We analysed questions raised by one group of students in group problem-solving and design process respectively. The number and proportion of student's questions are tabulated in Table 12.5. In the excerpts that follow, identifiable group members are represented by "G1" to "G4".

Clarification questions were the most prevalent among questions raised by students, taking up to 38.7% (Table 12.5). When clarifying, students often

Table 12.5 Type, number, and proportion of students' questions

Type of question	Number of student's questions	Proportion of total questions (%)
Generic task-procedural	12	19.4
Specific task-procedural	14	22.6
Clarification	24	38.7
Miscellaneous	12	19.4
Total	**62**	**100**

rephrased the previous utterance (Tan et al., 2017) or simply probed *"what is that?"* when referring to the sketch of the farming system. We noticed that two-thirds of the clarification questions were made during the design process when group members probed their peers as he drew the design of their aquaponics system. Some clarifications made were, *"Oh, you mean the fishes here?"*, *"Is that actually the water pump?"*, and *"Then how do they (crops) get the water?"*.

Excerpt 12.3 shows an exchange involving clarification questions where two members discussed the type of plants that can be planted using rooftop greenery as a farming method. It showed G3 seeking confirmation to register new information about the rooftop farming method that they have researched. The action of clarifying benefits the group in achieving a common understanding, eliminating confusion, and it also displays the thought processes of the students (Tan et al., 2017).

Task-procedural questions can be generic or specific depending on whether the student is asking, *"what should we do?"* in general or if the question is targeted at a particular information. Specific task-procedural questions often include conceptual knowledge or evidence in the question stem. For instance, one student first asked generically, *"what are we going to do?"* before rephrasing his question to be more targeted, *"what other types of agriculture are there in Singapore?"*. From Excerpt 12.4, it is observed that generic task-procedural questions may take the form of figuring out loud, showing the student's wonderment before realising the task that needs to be performed or arriving at a necessary conclusion. To add on, Excerpt 12.4 illustrates how students utilised specific task-procedural questions to weigh the benefits and limitations of the farming

Excerpt 12.3. You need to water it?

	Type of student's question
G3: *So, it depends on the rain?*	Clarification
G1: No, it doesn't really depend on the rain, cause it's drought tolerant. Like cactus lah.	
G3: *Oh, like do you need to water it?*	Clarification
G1: I mean ya of course. Plants need water.	
G3: *So, like if you water it then why must it be drought-tolerant?*	Clarification
G1: It's the it's the plant's characteristic.	

Excerpt 12.4. Just one?

	Type of student's question
G1: *So, we'll use this as the main. And then which one do you want to use? Rooftop greenery or hydroponics?*	Specific task-procedural
G1: You can take two if you want.	
G3: Nah, I don't think we should take two, just one is enough.	
G1: *Just one?*	Specific task-procedural
G3: Ya.	
G1: *Why?*	Specific task-procedural
G3: Taking two is going to get confusing.	
G1: You can use hydroponics also you know. *Because if the water isn't enough for the fishes, how is it going to supply back to the plant?*	Specific task-procedural

method that they have found. Thus, apart from providing insights about students' thinking processes, task-procedural questions also illustrated evidence of informed design such as balancing pros and cons (Purzer et al., 2015).

From the analysed video, we also observed several times when the students were distracted and went off-topic. During those times, their attention was brought back to the task when a group member (G1) reminded his peers to concentrate on the task. It should also be remarked that G1 led the discussion, drew the design, and appeared to be the most on-task. Since it would be difficult for the teacher to manage all groups simultaneously, a suggestion would be to assign a group leader in each group who will take charge of organising the team, structuring discussions, and ensuring that the members are on-task. This would also make group problem-solving more student-centred, as heavier responsibilities are placed on the students. Classroom management issues could also be reduced.

In conclusion, by analysing the teacher's questions, we identified Ally's appropriate use of framing questions and extensive use of Socratic questioning in facilitating students' understanding of the problem. From the student's questions, we noticed how students used clarification and task-procedural questions to foster discussion, guide their understanding, and construct knowledge. These are parallel to the functions of student's questioning as outlined by Chin and Osborne (2008). Thus, this study adds to the foundation of research on classroom interactions on problem-centric STEM activities.

Implications

As more economies invest resources to expose learners to integrated STEM learning, greater understanding of interactions among learners and between teachers and learners would enable better teaching strategies to be deployed. The use of complex, persistent, and extended problems in STEM activities demands time to be spent on helping learners make sense of the multiple factors involved.

As such, space must be created within STEM classrooms to enable learners to make sense of problems first before generating solutions.

Acknowledgements

This research is supported by grants from ERFP OER24/19 TAL and from Nanyang Technological University – URECA Undergraduate Research Programme. We would also like to express our appreciation for the teachers and students who actively participated in carrying out the STEM activity.

References

Barrett, B. S., Moran, A. L., & Woods, J. E. (2014). Meteorology meets engineering: An interdisciplinary STEM module for middle and early secondary school students. *International Journal of STEM Education*, *1*(6), 1–7. http://www.stemeducationjournal.com/content/1/1/7

Bereiter, C. (1992). Referent-centred and problem-centred knowledge. Elements of an educational epistemology. *Interchange*, *23*(4), 337–361.

Brown, R., Brown, J., Reardon, K., & Merrill, C. (2011). Understanding STEM: Current perceptions. *Technology and Engineering Teacher*, *70*(6), 5–9.

Bryan, L. A., Moore, T. J., Johnson, C. C., & Roehring, G. H. (2016). Integrated STEM education. In C. C. Johnson, E. E. Peters-Burton, & T. J. Moore (Eds.), *STEM Road Map: A framework for integrated STEM education* (pp. 23–38). New York: Routledge.

Chapin, S. H., O'Connor, C., & Anderson, N. C. (2009). *Classroom discussions: Using math talk to help students learn, Grades K-6*. Sausalito, CA: Math Solution Publications.

Chin, C., & Brown, D. E. (2002). Student-generated questions: A meaningful aspect of learning in science. *International Journal of Science Education*, *24*(5), 521–549.

Chin, C. (2006). Using self-questioning to promote pupils' process skills thinking. *School Science Review*, *87*(321), 113–122.

Chin, C. (2007). Teacher questioning in science classrooms: Approaches that stimulate productive thinking. *Journal of Research in Science Teaching*, *44*(6), 815–843.

Chin, C., & Osborne, J. (2008). Students' questions: A potential resource for teaching and learning science. *Studies in Science Education*, *44*(1), 1–39.

Chin, C., & Osborne, J. (2010). Supporting argumentation through students' questions: Case studies in science classrooms. *The Journal of the Learning Sciences*, *19*(2), 230–284.

Crismond, D. P., & Adams, R. S. (2012). The informed design teaching and learning matrix. *Journal of Engineering Education*, *101*(4), 738–797.

Dierdorp, A., Bakker, A., Van Maanen, J. A., & Eijkelhof, H. M. C. (2014). Meaningful statistics in professional practices as a bridge between mathematics and science: An evaluation of a design research project. *International Journal of STEM Education*, *1*(9), 1–15. http://www.stemeducationjournal.com/content/1/1/9

Duschl, R. (2008). Quality argumentation and epistemic criteria. In S. Erduran, & M. Aleixandre (Eds.), *Argumentation in Science Education* (pp. 159–175). Dordrecht, the Netherlands: Springer.

El-Deghaidy, H., & Mansour, N. (2015). Science teachers' perceptions of STEM education: Possibilities and challenges. *International Journal of Learning and Teaching*, *1*(1), 51–54.

English, L. D., & King, D. T. (2015). STEM learning through engineering design: Fourth-grade students' investigations in aerospace. *International Journal of STEM Education,* *2*(1), 14.

English, L. D., King, D., & Smeed, J. (2016). Advancing integrated STEM learning through engineering design: Sixth-grade students' design and construction of earthquake resistant buildings. *The Journal of Educational Research, 110*(3), 255–271.

Gillies, R. M., & Baffour, B. (2017). The effects of teacher-introduced multimodal representations and discourse on students' task engagement and scientific language during cooperative, inquiry-based science. *Instructional Science, 45*(4), 493–513.

Guzey, S. S., Moore, T. J., & Harwell, M. (2016). Building up STEM: An analysis of teacher-developed engineering design-based STEM integration curricular materials. *Journal of Pre-College Engineering Education Research, 6*(1), 1129.

Hardy, I., Kloetzer, B., Moeller, K., & Sodian, B. (2010). The analysis of classroom discourse: Elementary school science curricula advancing reasoning with evidence. *Educational Assessment, 15*(3–4), 197–221.

Hoey, M. (2001). *Textual Interaction: An Introduction to Written Discourse Analysis.* Portland: Psychology Press.

Honey, M., Pearson, G., & Schweingruber, H. (2014). *STEM Integration in K-12 Education: Status, Prospects, and Agenda for Research.* Washington, DC: National Academies Press.

Kuhn, D. (2009). Do students need to be taught how to reason? *Educational Research Review, 4*(1), 1–6.

Lee, S. C., & Irving, K. E. (2018). Development of two-dimensional classroom discourse analysis tool (CDAT): Scientific reasoning and dialog patterns in the secondary science classes. *International journal of STEM education, 5*(1), 5.

McDonald, C. V. (2016). A review of the contribution of the disciplines of science, technology, engineering and mathematics. *Science Education International, 27*(4), 530–569.

Michaels, S., & O'Connor, C. (2012). *Talk science primer. Cambridge, MA:* TERC.

Nam, Y., & Chen, Y. C. (2017). Promoting argumentative practice in socio-scientific issues through a science inquiry activity. *EURASIA Journal of Mathematics, Science and Technology Education, 13*(7), 3431–3461.

National Research Council (2012). *A framework for K-12 science education: Practices, crosscutting concepts, and core ideas.* Washington, DC.: The National Academies Press. https://doi.org/10.17226/13165.

Otto, S., Körner, F., Marschke, B. A., Merten, M. J., Brandt, S., Sotiriou, S., & Bogner, F. X. (2020). Deeper learning as integrated knowledge and fascination for science. *International Journal of Science Education, 1*(28), 807–834.

Paul, R., & Elder, L. (2007). Critical thinking: The art of Socratic questioning. *Journal of Developmental Education, 31*(1), 36–37.

Purzer, Ş., Goldstein, M. H., Adams, R. S., Xie, C., & Nourian, S. (2015). An exploratory study of informed engineering design behaviors associated with scientific explanations. *International Journal of STEM Education, 2*(1), 9.

Reeves, E. M. (2021). The need for STEM education: Now more than ever! *Southeast Asian Journal of STEM Education, 2*(1), 1–17.

Satchwell, R. E., & Loepp, F. L. (2002). Designing and implementing an integrated mathematics, science, and technology curriculum for the middle school. *Journal of Industrial Teacher Education, 39*(3), 41–66.

Slavit, D., Nelson, T. H., & Lesseig, K. (2016). The teachers' role in developing, opening, and nurturing an inclusive STEM-focused school. *International Journal of STEM Education, 3*(1), 1–17.

Star, J. R., Chen, J. A., Taylor, M. W., Durkin, K., Dede, C., & Chao, T. (2014). Studying technology-based strategies for enhancing motivation in mathematics. *International Journal of STEM Education*, *1*(7), 1–19. http://www.stemeducationjournal.com/content/1/1/7

Stohlmann, M., Moore, T. J., McClelland, J., & Roehrig, G. H. (2011). Impressions of a middle grades STEM integration program: Educators share lessons learned from the implementation of a middle grades STEM curriculum model. *Middle School Journal*, *1*, 32–40.

Tan, A. L., Lee, P. P. F., & Cheah, Y. H. (2017) Educating science teachers in the twenty-first century: Implications for pre-service teacher education. *Asia Pacific Journal of Education*, *37*(4), 453–471.

Tan, A. L., Teo, T. W., Choy, B. H., & Ong, Y. S. (2019). The S-T-E-M quartet. *Innovation and Education*, *1*(3), 1–14. doi: https://doi.org/10.1186/s42862-019-0005-x.

Thibaut, L., Ceuppens, S., De Loof, H., De Meester, J., Goovaerts, L., Struyf, A., Boeve-de Pauw, J., & Depaepe, F. (2018). Integrated STEM education: A systematic review of instructional practices in secondary education. *European Journal of STEM Education*, *3*(1), 02.

Toh, K. W. C., & Tan, A. L. (2020). *Development of interdisciplinary STEM curriculum planning and classroom observation protocols.* Paper presented at ASERA 2020 Online Conference, University of Wollongong, June 23–26, 2020.

United Nations. (2015). *Sustainable Developmental Goals.* Downloaded on January 6, 2021 from https://www.un.org/sustainabledevelopment/sustainable-development-goals/

Webb, N. M., Nemer, K. M., & Ing, M. (2006). Small-group reflections: Parallels between teacher discourse and student behavior in peer-directed groups. *The Journal of the Learning Sciences*, *15*(1), 63–119.

Wells, J. G. (2016). PIRPOSAL model of integrative STEM education: Conceptual and pedagogical framework for classroom implementation. *Technology and Engineering Teacher*, *75*(6), 12–19.

Winter, E. O. (1968). *Anaphoric sentence adjuncts. In Sentence and clause in scientific english.* In R. D. Huddleston, R. A., E. O., Winter, & A. Henrici (Eds.), Osti Project 5030. London, England: University College London.

World Economic Forum. (2020). Schools of the future: Defining new models of education for the fourth industrial revolution. Retrieved on Jan 18, 2020 from: https://www.weforum.org/reports/schools-of-the-future-defining-new-models-of-education-for-the-fourth-industrial-revolution/.

Contributor bios

Lay Ah-Nam is a Principal Assistant Director from the National STEM Centre, Ministry of Education, Malaysia.

Yu Chen, EdD, is a Research Associate at Department of Science and Environmental Studies and a Member of the Centre for Education in Environmental Sustainability of The Education University of Hong Kong.

Sherab Chophel is an in-service student at Samtse College of Education, currently pursuing his two years (2020–2021) program of Degree in Master of Education (MEd) in Biology under the full scholarship from Ministry of Education, Royal Government of Bhutan.

Ban Heng Choy is an Assistant Professor of Mathematics Education at the National Institute of Education, Nanyang Technological University, Singapore.

Hidayah Mohd Fadzil is a Senior Lecturer at the Faculty of Education, University of Malaya.

Chatree Faikhamta is an Associate Professor at the Division of Science Education, Department of Education, Faculty of Education, Kasetsart University, Thailand.

Lilia Halim is a Professor in Science Education in the STEM Education Enculturation Centre, Faculty of Education, Universiti Kebangsaan Malaysia.

Oksu Hong is currently a Director of the office of Science and Mathematics Education at Korea Foundation for the Advancement of Science and Creativity (KOFAC).

Roxanne Lau is a student-teacher at the National Institute of Education, Nanyang Technological University, Singapore.

Kornkanok Lertdechapat is a Lecturer at the Division of Curriculum and Instruction, Department of Curriculum and Instruction, Faculty of Education, Chulalongkorn University, Thailand

Auxencia A. Limjap is a Professorial Lecturer at the De La Salle University Science Education Dept. and currently the Research Director of Jose Rizal University.

Tian Luo is an Assistant Professor working at College of Teacher Education, Capital Normal University and a Member of the Centre for Education in Environmental Sustainability of The Education University of Hong Kong.

Ida Ah Chee Mok is an Associate Professor and former Associate Dean (2010–2021) in Faculty of Education at The University of Hong Kong.

Sheryl Lyn C. Monterola is a Professor at The University of the Philippines-Diliman, College of Education.

Gladys C. Nivera is a Professor in Mathematics Education at the Philippine Normal University, The National Center for Teacher Education in the Philippines.

Yann Shiou Ong is an Assistant Professor at the Natural Sciences and Science Education (Academic Group), National Institute of Education (NIE), Nanyang Technological University, Singapore.

Edwehna Elinore S. Paderna is an Associate Professor and current Chair of the Division of Curriculum and Instruction of the College of Education, University of the Philippines-Diliman.

Crist John M. Pastor is a full-time Faculty Member of the College of Graduate Studies and Teacher Education Research (CGSTER) and an Adjunct Faculty of The Faculty of Science, Technology, and Mathematics (FSTeM) of the Philippine Normal University.

Reeta Rai is a Lecturer and Coordinator for the STEM Education Research Centre (STEMRC) at the Samtse College of Education, Royal University of Bhutan.

Zhipeng Ren is a PhD student working in the area of Mathematics education at Faculty of Education, The University of Hong Kong.

Rohaida Mohd Saat is an Honorary Professor at the Department of Mathematics and Science Education, Faculty of Education, University of Malaya.

Edy Hafizan Mohd Shahali is a Senior Lecturer in the Department of Mathematics and Science Education, Faculty of Education, University of Malaya.

Winnie Wing Mui So is a Professor in the Department of Science and Environmental Studies and the Director of the Centre for Education in Environmental Sustainability of The Education University of Hong Kong as well as the Director of the Centre for Education in Environmental Sustainability.

Kinley is an Assistant Professor in the Department of Science Education at Samtse College of Education (SCE), Royal University of Bhutan (RUB).

Aik-Ling Tan is an Associate Professor and Deputy Head (Teaching & Curriculum Matters) at the Natural Sciences and Science Education (Academic Group) at the National Institute of Education, Nanyang Technological University, Singapore. She is also a core team member with the Multi-centric Education, Research and Industry STEM Centre at NIE (meriSTEM@NIE).

Paul Teng is the Managing Director at NIE International (NIEI), the education consulting company of Nanyang Technological University, Singapore.

Sherab Tenzin is a Teacher of Orong Central School (OCS) under Samdrup Jongkhar District, Ministry of Education (MoE) in Bhutan.

Tang Wee Teo, PhD, is an Associate Professor at the Natural Sciences and Science Education (Academic Group), National Institute of Education (NIE), Nanyang Technological University, Singapore. She is also the Co-Head of the Multi-centric Education, Research and Industry STEM Centre at NIE (meriSTEM@NIE).

Si Qi Toh is currently a student at the National Institute of Education Singapore, pursuing her Bachelors of Science (Education).

Tempa Wangchuk is a Teacher at Nganglam Central School, Pemagatshel District, Bhutan.

Index

For Product Safety Concerns and Information please contact our EU
representative GPSR@taylorandfrancis.com
Taylor & Francis Verlag GmbH, Kaufingerstraße 24, 80331 München, Germany

www.ingramcontent.com/pod-product-compliance
Lightning Source LLC
Chambersburg PA
CBHW060248220326
41598CB00027B/4028